Modern Systemic
Fungicides

Modern Systemic Fungicides

Dr. K.P. Singh

Oxford Book Company
Jaipur ▪ New Delhi

ISBN : 9789355246905

First published in India, 2026

Oxford Book Company
Regd. Off. "Bony Residency", Gate No. 2
Opp. Tilak Public School, Vishweswariya Nagar,
Gopalpura Road, Jaipur - 302018, India
Ph: 0141 - 2761280, Telefax: 2761381
email - oxfordbookjpr@gmail.com
website - oxfordbookcompany.com

Delhi Office: 102, 1st Floor, Satyam House,
4327/3, Ansari Road, Darya Ganj,
New Delhi - 110002, India
Ph: 011-45652440

Distributed By
KEYA INTERNATIONAL PRESS LLP
E-mail: keyainternationalpressllp@gmail.com

Printed and bound in India by Thomson Press Ltd., New Delhi

Contents

Preface

Fungicides are biocidal chemical compounds or biological organisms used to kill or inhibit fungi or fungal spores. Many fungicides are registered for use on plants, but only a few are readily available to gardeners. They are not restricted in use or categorized as highly toxic. Often they are difficult for the average gardener to obtain in small quantities. Understanding the disease cycle, proper timing, coverage and selection of the right fungicide are needed to get good control using fungicides. Many fungicides work by protecting healthy plant tissues.

Fungi can cause serious damage in agriculture, resulting in critical losses of yield, quality and profit. Fungicides are used both in agriculture and to fight fungal infections in animals. Chemicals used to control oomycetes, which are not fungi, are also referred to as fungicides as oomycetes use the same mechanisms as fungi to infect plants. Some of the fungicides, such as the sulfur and copper products, can be used for growing organic produce. Others, such as captan, triforine, and chlorothalonil, are synthetically produced and not used for organic gardening.

Fungicides can either be contact, translaminar or systemic. Contact fungicides are not taken up into the plant tissue, & only protect the plant where the spray is deposited; translaminar fungicides redistribute the fungicide from the upper, sprayed leaf surface to the lower, unsprayed surface; systemic fungicides are taken up & redistributed through the xylem vessels to the upper parts of the plant. New leaf growth is protected for a short period. The fungicide resistance action committee (FRAC) has several recommended practices to try to avoid the development of fungicide resistance, especially in at-risk fungicides including *Strobilurins* such as azoxystrobin.

Products should not be used in isolation but rather as mixture, or alternate sprays, with another fungicide with a different mechanism of action. The likelihood of the pathogen developing resistance is greatly decreased by the fact that any resistant isolates to one fungicide will hopefully be killed by the other – in other words two mutations would be required rather than just one. The effectiveness of this technique can be demonstrated by Metalaxyl, a phenylamide fungicide.

This book explores the latest advances in the field of this subject. The subject matter, both as regards the arrangement of chapters as well as contents is designed to meet the requirement of the students in several Universities.

— ***Dr. K.P. Singh***

Chapter 1

Fungicide

Fungicides are biocidal chemical compounds or biological organisms used to kill or inhibit fungi or fungal spores. Fungi can cause serious damage in agriculture, resulting in critical losses of yield, quality and profit. Fungicides are used both in agriculture and to fight fungal infections in animals. Chemicals used to control oomycetes, which are not fungi, are also referred to as fungicides as oomycetes use the same mechanisms as fungi to infect plants.

Fungicides can either be contact, translaminar or systemic. Contact fungicides are not taken up into the plant tissue, & only protect the plant where the spray is deposited; translaminar fungicides redistribute the fungicide from the upper, sprayed leaf surface to the lower, unsprayed surface; systemic fungicides are taken up & redistributed through the xylem vessels to the upper parts of the plant. New leaf growth is protected for a short period.

Most fungicides that can be bought retail are sold in a liquid form. A very common active ingredient is sulfur, present at 0.08% in weaker concentrates, and as high as 0.5% for more potent fungicides. Fungicides in powdered form are usually around 90% sulfur and are very toxic. Other active ingredients in fungicides include neem oil, rosemary oil, jojoba oil, the bacterium *Bacillus subtilis*, and the beneficial fungus *Ulocladium oudemansii*.

Fungicide residues have been found on food for human consumption, mostly from post-harvest treatments. Some fungicides are dangerous to human health, such as vinclozolin, which has now been removed from use. A number of fungicides are also used in human health care.

Resistance

Pathogens respond to the use of fungicides by evolving resistance. In the field several mechanisms of resistance have been identified. The evolution of fungicide resistance can be gradual or sudden. In qualitative or discrete resistance a mutation (normally to a single gene) produces a race of a fungus with a high degree of resistance.

Such resistant varieties also tend to show stability, persisting after the fungicide has been removed from the market. For example sugar beet leaf blotch remains resistant to azoles years after they were no longer used for control of the disease. This is because such mutations often have a high selection pressure when the fungicide is used, but there is low selection pressure to remove them in the absence of the fungicide.

In instances where resistance occurs more gradually a shift in sensitivity in the pathogen to the fungicide can be seen. Such resistance is polygenic – an accumulation of many mutation in different genes each having a small additive effect. This type of resistance is known as quantitative or continuous resistance. In this kind of resistance the pathogen population will revert back to a sensitive state if the fungicide is no longer applied.

Little is known about how variations in fungicide treatment affect the selection pressure to evolve resistance to that fungicide. Evidence shows that the doses that provide the most control of the disease also provide the largest selection pressure to acquire resistance, and that lower doses decreased the selection pressure.

In some cases when a pathogen evolves resistance to one fungicide it automatically obtains resistance to others – a phenomenon known as cross resistance. These additional fungicides are normally of the same chemical family or have the same mode of action, or can be detoxified by the same mechanism. Sometimes negative cross resistance occurs, where resistance to one chemical class of fungicides leads to an increase in sensitivity to a different chemical class of fungicides. This has been seen with carbendazim and diethofencarb.

There are also recorded incidences of pathogens evolving multiple drug resistance – resistance to two chemically different fungicides by separate mutation events. For example *Botrytis cinerea* is resistant to both azoles and dicarboximide fungicides.

There are several routes by which pathogens can evolve fungicide resistance. The most common mechanism appears to be alteration of the target site, particular as a defence against single site of action fungicides. For example Black Sigatoka, an economically important pathogen of banana, is resistant to the QoI fungicides, due to a single nucleotide change resulting one amino acid (glycine) being replaced by another (alanine) in the target protein of the QoI fungicides, cytochrome b. This presumably disrupts the binding of the fungicide to the protein, rendering the fungicide ineffective.

Upregulation of target genes can also render the fungicide ineffective. This is seen in DMI resistant strains of *Venturia inaequalis*.

Resistance to fungicides can also be developed by efficient efflux of the fungicide out of the cell. *Septoria tritici* has developed multiple drug resistance using this mechanism. The pathogen had 5 ABC type transporters with overlapping substrate specificities that together work to effectively pump toxic chemicals out of the cell.

In addition to the mechanisms outlined above, fungi may also develop metabolic pathways that circumvent the target protein, or acquire enzymes that enable metabolism of the fungicide to a harmless substance.

Fungicide Resistance Management

The fungicide resistance action committee (FRAC) has several recommended practices to try to avoid the development of fungicide resistance, especially in at-risk fungicides including *Strobilurins* such as azoxystrobin.

Products should not be used in isolation but rather as mixture, or alternate sprays, with another fungicide with a different mechanism of action. The likelihood of the pathogen developing resistance is greatly decreased by the fact that any resistant isolates to one fungicide will hopefully be killed by the other – in other words two mutations would be required rather than just one. The effectiveness of this technique can be demonstrated by Metalaxyl, a phenylamide fungicide. When used as the sole product in Ireland to control potato blight (*Phytophthora infestans*) resistance developed within one growing season. However in countries like the UK where it was only ever marketed as a mixture resistance problems developed more slowly.

Fungicides should only be applied when absolutely necessary, especially if they are in an at-risk group. Lowering the amount of fungicide in the environment lowers the selection pressure for resistance to develop. Manufacturers' doses should always be followed. These doses are normally designed to give the right balance between controlling the disease and limiting the risk of resistance development. Higher doses increase the selection pressure for single site mutations that confer resistance, as all strains but those that carry the mutation will be eliminated, and thus the resistant strain will propagate. Lower doses greatly increase the risk of polygenic resistance, as strains that are slightly less sensitive to the fungicide may survive.Template:Citation needed/date=Oct 2012

It is also recommended that where possible fungicides are only used in a protective manner, rather than to try to cure already infected crops. Far fewer fungicides have curative/ eradicative ability than protectant. Thus fungicide preparations advertised as having curative action may only have one active chemical; a single fungicide acting in isolation increases the risk of fungicide resistance.

It is better to use an integrative pest management approach to disease control, rather than relying on fungicides alone. This involves the use of resistant varieties and hygienic practises, such as the removal of potato discard piles and stubble on which the pathogen can overwinter, greatly reducing the titre of the pathogen and thus the risk of fungicide resistance development.

A Short History of Fungicides

A century ago the science of plant pathology was just being born and L. R. Jones wondered, at a meeting in Atlanta in 1913, whether it was a good thing to separate plant pathology as a discipline from botany (8). This was only a few decades after the idea that fungi were causal agents of plant diseases had been accepted and that the concept of "spontaneous generation," which regarded fungi as symptoms on plants suffering from bad environmental conditions, had been on its way out. This new concept of germs being causal agents of disease had its earliest supporters among plant pathologists like Tillet for bunt in 1755, Prevost for smut in 1807, Berkeley and de Bary for the late blight fungus of potatoes in the 1850s (13). Unfortunately, the broader significance of these findings was not recognized, and it was only in the 1860s that the work of Pasteur and Koch on human pathogens led to the break through of the germ theory (9). Therefore, it is not surprising that the understanding and the standard of fungicides a century ago was very rudimentary.

As with many inventions, "development" of the first fungicide was the result of good observations. The first use of brining of grain with salt water followed by liming took place in the middle of the 17th century to control bunt, and followed the observation that seed wheat salvaged from the sea was free of bunt. This had occurred long before Tillet (1755) established that seed-borne fungi (*Tilletia tritici*, *T. laevis*) caused bunt of wheat and that it could be controlled by seed treatments of lime, or lime and salt. Another important discovery was made in France in 1882 by Millardet, who noticed that grape vines that had been sprayed with a bluish-white mixture of copper sulfate and lime to deter pilferers retained their leaves through the season, whereas the unsprayed vines lost their leaves. After numerous spraying experiments Millardet concluded that a mixture of copper sulfate and hydrated lime could effectively control downy mildew of grape. Hooker in 1923 stated in his paper on colloidal copper hydroxide that there were no entirely satisfactory fungicides available. Then he went on to describe the shortcomings of the Bordeaux mixture and lime-sulfur, the latter also being "most disagreeable to handle." Up until the 1940s chemical disease control relied upon inorganic chemical preparations, frequently prepared by the user.

Many of the early efforts to produce healthy crops involved diseases that had newly been introduced and left the growers quite helpless. In the following sections we highlight trends in fungicide development and use over the last century in light of the ever-changing spectra and intensities of fungal pathogens, which have often occurred as a consequence of changing cropping systems.

Development of the Importance of Diseases

Since Tillet presented the results of well replicated and controlled experiments in 1755, in which he added black dust from bunted wheat to seed from healthy wheat and observed that bunt was much more prevalent in plants produced from such seed than from non-dusted seed, researchers and growers have been combating plant pathogens in various ways. The importance of epidemics was highlighted by late blight of potato in Europe in the 1840s which led to several severe famines.

Key among the researchers at this time was DeBary, whose studies on the Peronosporaceae and the discovery of alternate hosts in the life cycle of rust fungi were two of his many important contributions. At that time it took 100 years for man's knowledge to double, today it is done in 16 years. In the following 150 years much has been learnt about the control of plant diseases and several complementary approaches were developed for their control (Table 1). Depending on the crop, the disease and the availability of control methods, a different set of approaches is employed.

Table: *Key methods of controlling plant diseases.*

Regulatory measures	Quarantine, inspection and seed certification
Cultural methods	Crop rotation, sanitation, improved growing conditions
Biological methods	Breeding of resistant varieties, microbials
Chemical control	Seed treatment; soil, foliar & post harvest applications

When looking at disease losses, the most devastating ones occur post harvest, as these include the entire costs incurred in producing the crop. One of the greatest discoveries in plant pathology in the twentieth century was that of Flor in the 1940s when he proposed his gene for gene theory, based on results of his crosses of flax, each of which were resistant to none, one, or several races of flax rust (3).

Flor found that both resistance in plants as well as avirulence in the pathogen were inherited, and that both traits were dominant. Plant breeders at universities and seed companies have invested thousands of man-years in breeding crops resistant to diseases since that time. It may take as long to breed a new resistance gene into a new cultivar as to introduce a new fungicide (± 10 years), even though this time can be shortened somewhat today with molecular tools by reducing the back-crossing cycles. Pathogens may overcome both means of control by the selection of adapted strains in a short period of time, thereby invoking the old adage that *nature abhors a vacuum.*

More recently, the trend towards more intensive cropping, the introduction of high yielding varieties and the neglect of proper crop rotation have led to new diseases in some crops. Minimum tillage or direct seeding without any plowing as well as mechanization of harvesting in fruit trees or grapes can also change the spectrum of diseases for which the growers need to find control measures.

Such changes in crop varieties and in production methods keep the disease picture very dynamic so that new methods for disease control are constantly needed.

Overview of Fungicide Development and Usage

Up to 1940. Through the study of diseases that caused clear economic damage and the study of epidemiology of the pathogens, the basic principles of disease control were established.

The major products used up to 1940 are listed in Table 2. In general, chemical disease control was aimed at horticultural crops (fruit and vegetables) as well as seed treatments. Concerns for the products' impact on the environment were largely non-existent, as were concerns for the applicator. Most users prepared their own fungicides from basic recipes.

***Table:** Fungicides in use up until 1940 [after Russell 2005 (15)]*

Year	*Fungicide*	*Primary Use*
1637	Brine	Cereal seed treatment
1755	Arsenic	Cereal seed treatment
1760	Copper sulfate	Cereal seed treatment
1824	Sulfur (dust)	Powdery mildew and other pathogens
1833	Lime sulfur	Broad spectrum foliar pathogens
1885	Bordeaux mixture	Broad spectrum foliar pathogens
1891	Mercury chloride	Turf fungicide
1900	$CuOCl_2$	Especially *Phytophthora infestans*

1914	Phenylmercury chloride	Cereal seed treatment
1932	Cu_2O	Seed and broad spectrum foliar diseases
1934	Dithiocarbamates patented	Broad spectrum protectants
1940	Chloranil, Dichlone	Broad spectrum seed treatment

Proprietary products were available at this time for those who did not want to take the time and trouble to make their own. Prior to the introduction of the dithiocarbamates, the testing of which was described by McCallan of Cornell University in 1930, most of the products used as fungicides were applied at high rates, e.g., 10 to 20 kg a.i./ha (~9 to 18 lbs/acre) for sulfur against powdery mildew on grapes. The products did not always give good control, could be phytotoxic and had to be applied frequently. This is a reflection of the drastic reduction of use rates per ha (acre) as more effective and selective fungicides were introduced over this time. For example, the current use rates of well below 100 gr/ha (1.4 oz/acre) for many triazoles against the same pathogen is a 200 fold reduction.

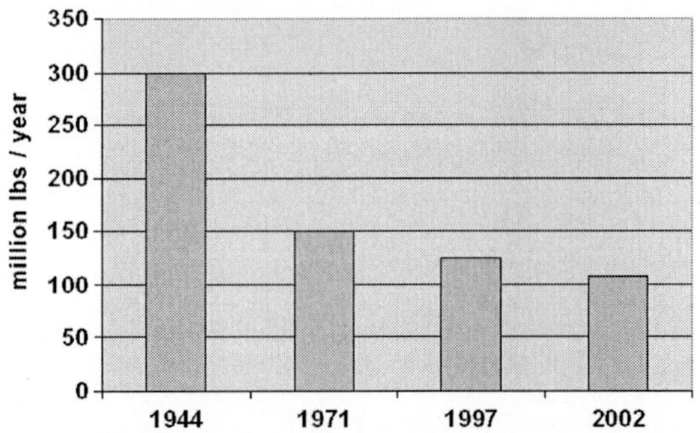

Figure: *U.S. Crop Protection Fungicide Use (5).*

1940 to 1970. From 1940 to 1970 there were a number of new chemistry classes introduced as fungicides (Table 3). The dithiocarbamates and later the phthalimides represented a major improvement over the previously used inorganic fungicides in that they were more active, less phytotoxic and easier to prepare by the user. The fungicide research climate of that time is well described by Horsfall (7), who was instrumental in the discovery and the introduction of the dithiocarbamates, the most widely used group of organic fungicides.

Table: *Key classes of fungicides introduced between 1940 and 1970.*

Fungicide Class	*Active Ingredient and Year*
Dithiocarbamate	thiram 1942, zineb, nabam 1943, maneb 1955, mancozeb 1961
Aromatic Hydrocarbon	biphenyl 1944
Phthalimide	captan, folpet 1952, captafol 1962
Fentin	fentin acetate, fentin hydroxide 1954
Antibiotic	blasticidin S 1955, kasugamycin, polyoxin 1965
Triazine	anilazine 1955
Guanidine	dodine 1957
Nitroanaline	dicloran 1960
Benzimidazoles	thiabendazole 1964, benomyl 1968, thiophanate methyl 1970
Phthalonitrile	chlorothalonil 1964
Morpholine	dodemorph 1965, tridemorph 1969
Carboxanilide	Carboxin 1966, oxycarboxin 1966

It is Worth Noting that During this Time Period: (i) several of these classes of chemistry also gave rise to herbicides — e.g., triazines and nitroanalines, (ii) an antibiotic was introduced to control rice blast in Japan and is classified as a systemic and, (iii) the swing had taken place to the reliance on commercially produced products helping to give rise to the crop protection industry. With this swing came the interest in the biochemical mode of action of fungicides. Most of these fungicides were being used at 1.5 to 3 kg a.i./ha (~1.3 to 2.7 lb/acre).

The decade from 1960 to 1970 saw a rapid expansion of research and development along with a rapid growth of the fungicide markets. In this decade, the most widely used protectant fungicides, mancozeb and chlorothalonil, were introduced. The decade also gave us the first broad-spectrum foliar systemic, thiabendazole, and the systemic seed treatment carboxin. Much of this new chemistry arose as a result of the basic manufacturers having moved away from *in vitro* screens to *in vivo* screens. In these *in vivo* screens, young potted plants were sprayed or drench treated with the test compounds and inoculated with a major pathogen soon thereafter. The plants then were placed in a greenhouse or growth chamber under ideal disease conditions until disease symptoms appeared and they could be rated for contact or systemic efficacy as well as for phytotoxicity. This process was usually repeated on a weekly basis with large numbers of test compounds.

Beyond 1970: The more important modern fungicides introduced since 1970 are listed in Table 4 according to their mode of action or chemical class. For additional information on mode of action and resistance risk. A discussion of the key classes of modern fungicides follows.

Benzimidazole: fungicides were introduced for plant disease control in the 1960s and early 1970s as foliar fungicides, seed treatments and for use in post harvest applications. They possessed unique properties not seen before in the protectants. These included low use rates, broad spectrum and systemicity with post-infection action that allowed for extended spray interval. All these qualities made them very popular with growers but also subject to misuse, such as poor spray coverage and curative spraying. The first case of resistance to benzimidazoles occurred in powdery mildew in greenhouses in 1969, one year after introduction. By 1984, resistance had been reported on many of the pathogens against which benzimidazoles are active (17). The reason for the rapid development of resistance was that these fungicides were single site inhibitors of fungal microtubule assembly during mitosis, via tubulin-benzimidazole-interactions. The primary patent holders of this class were DuPont (Benlate), Merck, Sharp & Dohme (Mertec) and Nippon Soda (Topsin M). The current ranking of global sales is: carbendazim, thiophanate, thiabendazole.

Morpholine: fungicides are best known for their excellent control of cereal diseases, powdery mildew on vegetables and grapes, and sigatoka of banana. During the 1980s fenpropidin and fenpropimorph were key fungicides in the European cereal market, while tridemorph was used extensively for sigatoka. This class of chemistry, although having seen shifts in sensitivity by some pathogens (sigatoka in Central and South America), is still in use. Key patents were held by BASF (Calixin and Corbel) and Dr. R. Maag (Corbel and Tern). Dimethomorph, though a morpholine, is quite distinct from the morpholines above with its activity against Oomycetes via the inhibition of cell wall formation (FRAC group 40 in Table 5). The current ranking of global sales is: dimethomorph, fenpropidin, fnpropimorph, sprioxamine.

Morpholine fungicides belong to a broad group of fungicides that is often referred to as sterol biosynthesis inhibitors (SBI). Other SBIs include the next four groups of fungicides.

***Piperazines*:** the major player in this group was triforine, which was used extensively as a home and garden product (especially on roses). Key to the acceptance of triforine was its efficacy and safety to a wide range of plants. The key producer was CelaMerck (Saprol).

Imidazoles: include a small number of compounds in this class that are active against plant pathogens. The most important are imazalil (Janssen) and prochloraz (Boots). The primary uses for imazalil were as a seed treatment and post harvest treatment, while prochloraz (Sportak) was used on cereals, being especially active on *Pseudocercosporella* eyespot.

Pyrimidines: are a class of fungicides that were extensively explored by Eli Lilly giving rise to nuarimol, fenarimol and triarimol. The major player of these was fenarimol (Rubigan) on pome fruit, grapes and turf.

Triazoles: are the largest class of fungicides. Bayer was the first to launch a triazole, namely triadimefon (Bayleton) in 1973. This was soon followed by triadimenol (Baytan) and bitertanol (Baycor). Janssen Pharmaceuticals sold the agricultural use rights to Ciba-Geigy for propiconazole (Tilt) which was launched in 1979. Numerous other triazoles have been launched since, with Bayer's most recent entrée being prothiaconazole (Proline) in 2004. The reason for the longevity of this class of fungicides is that while being highly efficient broad spectrum products, resistance has occurred over time as a slow shift resulting in a decreased sensitivity to their mode of action as de-methylation inhibitors (DMI). The newer triazoles, being intrinsically more active, push the sensitivity curves back to their original ED 50 values. The current ranking of global sales is: tebuconazole, epoxiconazole, propiconazole, difenoconazole, flusilazole, tetraconazole, fluquinconazole, flutriafol.

Anilides: are a diverse group of fungicides. The earliest introduction was anilazine (Dyrene), primarily as a leaf spot fungicide from Bayer and Nissan, followed by the seed treatment carboxin (Vitavax), which is highly effective on bunts, smuts and assorted Basidiomycetes such as *Rhizoctonia spp*. This was followed by the dicarboximides iprodione (Rovral) from Rhone-Poulenc, vinclozolin (Ronilan) from BASF and procymidone (Sumisclex) from Sumitomo. These fungicides all had exceptional protectant activity on the genera *Botrytis, Monilinia* and *Sclerotinia*. Combating resistance became an issue with the wide scale use of these fungicides.

Unquestionably the greatest of this group of anilides were the phenylamide fungicides metalaxyl (Apron/ Ridomil) from Ciba–Geigy and benalaxyl (Galben) from Isagro. These, along with phosphonate fosetyl-Al (Aliette) from Rhone-Poulenc, which was also introduced in 1977, brought a completely new level of control to the Oomycetes through their systemic properties by offering protection to the plants as seed treatments, and soil or foliar applications. Oxadixyl (Sandofan) from Sandoz was a later member of the phenylamides. Recently, Syngenta (1996) with mefenoxam (Apron XL and Ridomil Gold) and Isagro (2005) with kiralaxyl have introduced the resolved isomers of metalaxyl and benalaxyl. Again, what has limited the use of the phenylamide fungicides has been the development of resistance, even though the manufacturers tried introducing combinations with protectant fungicides such as mancozeb and chlorothalonil. The latest anilide to be registered (2003) is boscalid (Emerald, Endura, Pristine) from BASF. Boscalid is registered for foliar use on a wide range of vegetables, fruits and nut crops, either alone or in a mixture with pyraclostrobin as Pristine.

Strobilurins:, launched in 1996, are now the second largest chemistry group of fungicides as a result of widespread use on cereals and, more recently, on soybeans (a market that reached $600 million in 2004). Companies have recently also promoted the plant health attributes of this group of fungicides on soybeans and corn. The strobilurin fungicides have a broad spectrum, are highly efficacious, and are suitable for a wide range of crops. Some problems with disease resistance are affecting sales (e.g., *Septoria* in wheat in Europe, and the U.S. turf market). As a result, companies are adjusting the use recommendations by developing mixtures and other uses, including seed treatments.

carboxylic Acid Amides

Zeneca began researching this chemistry in the early 1980s, first synthesizing azoxystrobin (Amistar, Abound, Quadris) in 1988; it is now the largest selling member of this group. However, kresoxim-methyl (Cygnus, Sovran) from BASF was the first member to be commercialized in 1996. BASF has since entered the market (2002) with a broader spectrum strobilurin pyraclostrobin (Cabrio, Headline, Insignia) also sold in combination with kresoxim-methyl as Opera and with boscalid as Pristine.

Other strobilurin fungicides include, trifloxystrobin (Flint) discovered by MAAG but sold by Bayer, and fluoxastrobin (Disarm, Evito) discovered by Bayer but sold by Arysta.

The current ranking of global sales is: azoxystrobin, pyraclostrobin, trifloxystrobin, kresoxim-methyl, picoxystrobin.

Other Systemic Fungicides include a diverse group of products, such as: tricyclazole (Beam) launched in 1975 by Eli Lily/Dow and still widely used for control of rice blast; cymoxanil (Curzate), a downy mildewcide from DuPont; the cereal and fruit fungicide cyprodinil (Vanguard, Unix) from Syngenta; fludioxonil (Saphire, Switch, Maxim) from Syngenta; and quinoxyfen (Fortress, Quintec), a powdery mildewcide from Dow.

Newer active ingredients introduced recently are benthiavalicarb (Valbon from Kumiai) and mandipropamid (Revus from Syngenta), from the carboxylic acid amide (CAA) fungicide group, and fluopicolide (Infinito from Bayer), metrafenone (Flexity from BASF), proquinazid (Talius from DuPont), and zoxamide (Electis from Dow).

For a more detailed technical description of modern fungicides, including chemistry, we refer to the treatise edited by Krämer and Schirmer (10).

One of the most novel new products introduced by Ciba-Geigy is acibenzolar-S-methyl (Actigard, Bion). At use rates of 30 gr/ha (~0.4 oz/acre) or less, it activates the host's systemic acquired resistance (SAR) process in many crop plants.

It offers broad protection against fungi, bacteria and viruses without having any direct activity on these pathogens (11). Actigard has performed best when incorporated into a program of chemical sprays, as the inherent level of disease control has seldom been sufficient when applied alone. This product has initiated a whole new field of research into utilizing peptides for controlling diseases, and other means of stimulating SAR and the Jasmonic acid pathway (JA) with chemicals and biological agents in plants. Probenazole has been used successfully against rice blast since 1979 and was later shown to activate defense mechanisms in rice.

Fungicide Resistance Management

In Figure depicted, market shares for the major chemical groups of fungicides are summarized for 2005 including non-crop uses. According to this source, the youngest group, the stobilurines, had surpassed within a few years all of the older fungicides groups in importance except only the DMIs. This indicates that the innovation potential has been substantial for fungicides with new modes of action and excellent performance, not least because some of the older groups have suffered from an erosion of efficacy against some important pathogens due to the emergence of resistance. The non-crop uses include amenity grass, which is the single largest fungicide market in the U.S. The numerous pathogens that attack these highly maintained grasses, such as those found on golf courses, frequently require weekly spray applications through out the summer. This has already led to resistance development in *Magnaporthe grisea* (gray leaf spot) to strobilurins.

Thus, a major consideration for the design of fungicide use strategies is the threat of fungicide resistance. There have been considerable efforts by industry to conduct research in the areas of mode of action, resistance risk, field monitoring for baseline sensitivity and sensitivity variations in treated fields. Based on the results, use strategies are designed that reduce the risk of fungicide resistance build-up or worse, the loss of efficacy of whole fungicide classes.

This threat of fungicide resistance and the fact that cross-resistance often exists to related products from different manufacturers has lead to a close collaboration between them in FRAC (Fungicide Resistance Action Committee). Results from research in mode of action, resistance risk and field monitoring are pooled and strategies are developed to minimize the risk of resistance build-up. FRAC has produced several monographs on various aspects of fungicide resistance and has grouped the available fungicides according to various criteria that facilitate the understanding of the resistance risk of the different fungicide groups (Table 5). Such assessments of the resistance risk are made difficult by the unpredictability of cross-resistance, which, in most cases, is clear-cut and follows the mode of action but in other cases can be quite complex. In the end, what matters for practical purposes, is not so much the mode of action of a fungicide but its cross-resistance to other fungicides.

Table: *Mode of action of major fungicides classes, their FRAC code and resistance risk.*

FRAC Code	***Chemical Class***	***Mode of action / inhibition***	***Resistance risk***
1	Benzimidazoles	Beta-tubulin biosynthesis	high
2	Dicarboximides	NADH cytochrome c reductase in lipids	high
3	Azoles, Pyrimidines	C-14 demethylation in sterol biosynthesis	medium
4	Phenylamides	RNA polymerase	high
5	Morpholines	^8 and ^7 isomerase and ^14 reductase in sterol biosynthesis	low-medium
7	Carboxamides	Succinic acid oxidation	medium
9	Anilinopyrimidine	Methionine biosynthesis	medium
11	Strobilurins	Mitochondrial synthesis in cytochrome bc1	high
16	Various chemistry	Melanin biosynthesis (two sites)	medium
40	Carboxylic acid amides	Cell wall formation in Oomycetes	low-medium
M1	Inorganics	Multisite contact	low
M3	Dithiocarbamates	Multisite contact	low
M5	Phthalimides	Multisite contact	low

The Benefits and Risks of Fungicides

The application of any chemical to a crop or food raises the question of risks and benefits. This discussion of risk has shifted from dealing with the toxicity to the user in the field and the consumer (4) to a much wider focus that includes the whole environment and the ecosystem in which the crops are growing (16). As a consequence more and more studies are required before a fungicide can be used, leading to enormous development costs. This leads industry to concentrate on the big markets, while smaller markets are increasingly left out and in urgent need of effective fungicides. In the U.S. the IR-4 program has been established to provide safe and effective pest management solutions for speciality crop growers. In December 2007 the UN-FAO held a global minor use summit along with IR-4 and EPA to establish global residue zones and standard data requirements.

Overall, most analyses come to the conclusion that the benefits of fungicides far outweigh the risks, if they are used carefully and according to the label recommendations. Currently more than 80% of the fruit and vegetable crops grown in the U.S. receive a fungicide every season.

The benefit of fungicide use in the U.S. agriculture is said by Gianessi and Reigner (5) to boost farm income by nearly $13 billion annually. The alternatives proposed by organic farmers, who are opposed to intensive farming altogether, exclude the use of synthetic fungicides, but allow the use of copper and sulfur based inorganic fungicides. There is still an ongoing debate as to whether traditionally or organically grown products are safer for the consumer. For example, a growing number of studies are being conducted to evaluate the risk of mycotoxins in the two farming systems.

The Future of Fungicides

A look at the history of fungicides should give us some idea of what to expect in the future. The major changes in fungicide use have usually been associated with changes in the spectra of pathogens as well as in crop intensities, practices or prices. The migration of tobacco blue mold into Europe or soybean rust into the Americas had a dramatic impact on the fungicide use on these crops.

Figure shows the value of fungicides of the major crop groups and illustrates the importance of the dynamic nature of the fungicide market. Cereals, especially the foliar applications, represent a relatively recent market when compared to fruits and vegetables. The soybean market was negligible until a few years ago when rust reached the large soybean acreages in North and South America. Such shifts in pathogen spectra could not be foreseen and will continue to occur in the future due to increased global trade with plant material. Similarly the dramatic increase in cropping intensity in European cereal production in the 1960s and 70s created a major market where practically none existed before. The remarkable ability of new pathogens to adapt to intensively cultivated cereals has led to a large list of pathogens that can threaten these crops. More generally, a growing world population that wants to be fed better than today will lead to increased areas of intensive cropping and hence, most likely, fungicide use will also increase. The higher demand on some grain commodities for use as biofuel has already resulted in very low carryover stocks of grain, which will inevitably result in price increases (2). This exerts pressure for further intensification of these crops. Corn in the U.S. is a good candidate for fungicide sprays becoming a regular production practice.

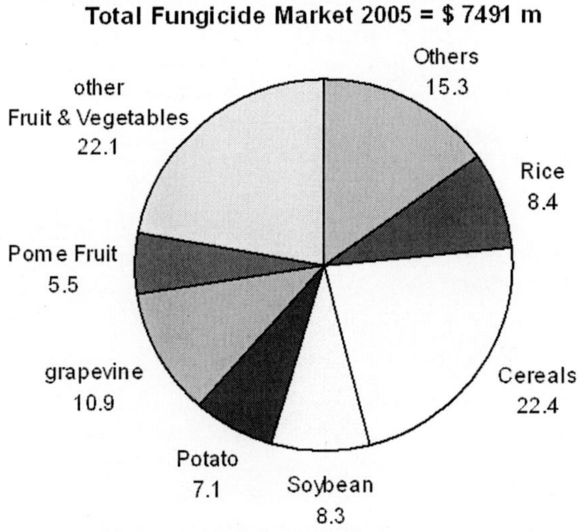

Figure: Global fungicides market share for the major crop groups for 2005 (14).

The trend towards a more judicious use of fungicides in conjunction with disease forecast systems that has been observed in the past can be expected to continue in the future. This will help reduce the risk of adaptation by the target fungi and at the same time will reduce residues in the environment and on the produce. The efforts of breeding for disease resistance will also continue and possibly increase by utilizing the tools of genetic engineering. Both genetic resistance and selective fungicides are prone to adaptation by the pathogen. The balance between genetic and chemical control will therefore most likely continue and research in both areas will complement each other to assure the availability of effective combinations of host resistance and fungicides for crops that should produce ever higher yields of ever better quality.

Currently the biocontrol products contribute less than 1% of the fungicide market; undoubtedly they will contribute more in the future, with the most likely avenue being through natural compounds known as biopesticides. An interesting new area of research is the use of antimicrobial peptides (AMP) for improving resistance to pathogens using transgenic plants as bio-factories for fungicides or bactericides.

Thus the major multi-national companies will be focusing their future on an integration of genetic traits and agrochemicals. Projected growth of the agrochemical market is 3.4% over the next 5 years, whereas traits are forecast to grow at 7%. The average current (2005) R & D expenditures of the six major Agro companies is 64.5% for chemistry and 35.5% for seed and traits. This ensures that new fungicides will continue to be developed to protect the ever more precious cultivars, where they do not have sufficient genetic disease resistance. This balance between genetic resistance and disease control products of chemical, biochemical or biological nature will remain and is not likely to change dramatically in the near future. Equally important for sustainable disease control will be the intelligent integration of these technologies with sound cultural and sanitation measures.

Benzimidazole

Benzimidazole is a heterocyclic aromatic organic compound. This bicyclic compound consists of the fusion of benzene and imidazole. The most prominent benzimidazole compound in nature is *N*-ribosyl-dimethylbenzimidazole, which serves as an axial ligand for cobalt in vitamin B_{12}.

Benzimidazole, in an extension of the well-elaborated imidazole system, has been used as carbon skeletons for N-heterocyclic carbenes. The NHCs are usually used as ligands for transition metal complexes. They are often prepared by deprotonating an N,N'-disubstituted benzimidazolium salt at the 2-position with a base.

Preparation

Benzimidazole is commercially available. The usual synthesis involves condensation of o-phenylenediamine with formic acid, or the equivalent trimethyl orthoformate:

$$C_6H_4(NH_2)_2 + HC(OCH_3)_3 \rightarrow C_6H_4N(NH)CH + 3\ CH_3OH$$

By altering the carboxylic acid used, this method is generally able to afford 2-substituted benzimidazoles. Benzimidazole also has fungicidal properties. It acts by binding to the fungal microtubules and stopping hyphal growth. It also binds to the spindle microtubules and blocks nuclear division.

Carbamate

Carbamates are organic compounds derived from carbamic acid (NH_2COOH). A carbamate group, carbamate ester (e.g., ethyl carbamate), and carbamic acids are functional groups that are inter-related structurally and often are interconverted chemically. Carbamate esters are also called urethanes.

Synthesis

Carbamic acids are derived from amines:

$$R_2NH + CO_2 \rightarrow R_2NCO_2H$$

Carbamic acid is about as acidic as acetic acid. Ionization of a proton gives the carbamate anion, the conjugate base of carbamic acid:

$$R_2NCO_2H \rightarrow R_2NCO_2^- + H^+$$

Carbamates also arise via hydrolysis of chloroformamides:

$$R_2NC(O)Cl + H_2O \rightarrow R_2NCO_2H + HCl$$

Carbamates may be formed from the Curtius Rearrangement, where isocyanates formed are reacted with an alcohol.

$$RNCO + R'OH \rightarrow RNHCO_2R'$$

Applications and Occurrence

Although most of this article concerns organic carbamates, the inorganic salt ammonium carbamate is produced on a large scale as an intermediate in the production of the commodity chemical urea from ammonia and carbon dioxide.

Carbamates in Biochemistry

N-terminal amino groups of valine residues in the á- and â-chains of deoxyhemoglobin exist as carbamates. They help to stabilise the protein, when it becomes deoxyhemoglobin and increases the likelihood of the release of remaining oxygen molecules bound to the protein. The influence of these carbamates on the affinity of hemoglobin for O_2 is called the Bohr effect. The å-amino groups of the lysine residues in urease and phosphotriesterase also feature carbamate. The carbamate derived from aminoimidazole is an intermediate in the biosynthesis of inosine. Carbamoyl phosphate is generated from carboxyphosphate rather than CO_2.

CO_2 Capture by Ribulose 1,5-Bisphosphate Carboxylase

Perhaps the most important carbamate is the one involved in the capture of CO_2 by plants since this process is relevant to global warming. The enzyme Ribulose 1,5-bisphosphate carboxylase/oxygenase fixes a molecule of carbon dioxide as a carbamate at the start of the Calvin cycle). At the active site of the enzyme, a Mg^{2+} ion is bound to glutamate and aspartate residues as well as a lysine carbamate. The carbamate is formed when an uncharged lysine side-chain near the ion reacts with a carbon dioxide molecule from the air (*not* the substrate carbon dioxide molecule), which then renders it charged, and, therefore, able to bind the Mg^{2+} ion.

Commercial Carbamate Compounds

Carbamate Insecticides

The so-called carbamate insecticides feature the carbamate ester functional group. Included in this group are aldicarb, carbofuran (Furadan), carbaryl (Sevin), ethienocarb, fenobucarb, oxamyl and methomyl. These insecticides kill insects by reversibly inactivating the enzyme acetylcholinesterase. The organophosphate pesticides also inhibit this enzyme, although irreversibly, and cause a more severe form of cholinergic poisoning. Fenoxycarb has a carbamate group but acts as a juvenile hormone mimic, rather than inactivating acetylcholinesterase. The insect repellent icaridin is a substituted carbamate.

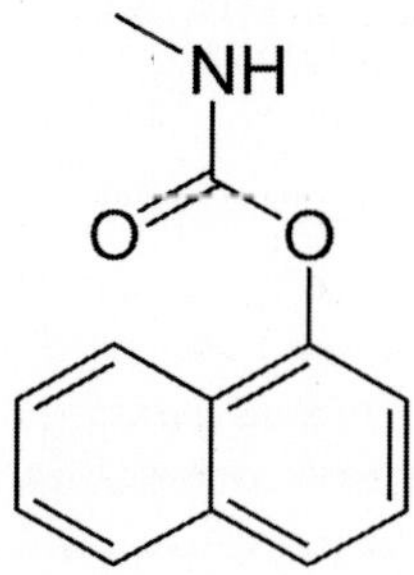

Figure: The carbamate insecticide Carbaryl.

Polyurethanes

Polyurethanes contain multiple carbamate groups as part of their structure. The "urethane" in the name "polyurethane" refers to these carbamate groups; ethyl carbamate (common name "urethane") is neither a component of polyurethanes, nor is used in their manufacture. Polyurethane polymers have a wide range of properties and are commercially available as foams, elastomers, and solids. Typically, polyurethane polymers are made by combining diisocyanates, e.g. toluene diisocyanate, and diols, where the carbamate groups are formed by reaction of the alcohols with the isocyanates:

$$RN{=}C{=}O + R'OH \rightarrow RNHC(O)OR'$$

Preservatives and Cosmetics

Iodopropynyl butylcarbamate is a wood and paint preservative and used in cosmetics.

In Human Medicine

Urethane or ethyl carbamate was once produced commercially in the United States as an antineoplastic agent and for other medicinal purposes. It was found to be toxic and largely ineffective. It is occasionally used as a veterinary medicine. In addition, some carbamates are used in human pharmacotherapy, for example, the cholinesterase inhibitors neostigmine and rivastigmine, whose chemical structure is based on the natural alkaloid physostigmine. Other examples are meprobamate and its derivatives like carisoprodol, felbamate, and tybamate, a class of anxiolytic and muscle relaxant drugs widely used in the 60s before the rise of benzodiazepines, and still used nowadays in some cases. The protease inhibitor darunavir for HIV treatment also contains a carbamate functional group.

Sulfur Analogues

There are two oxygen atoms in a carbamate (1), $ROC(=O)NR_2$, and either or both of them can be conceptually replaced by sulfur. Analogues of carbamates with only one of the oxygens replaced by sulfur are called thiocarbamates (2 and 3). Carbamates with both oxygens replaced by sulfur are called dithiocarbamates (4), $RSC(=S)NR_2$.

There are two different structurally isomeric types of thiocarbamate:

- *O*-thiocarbamates (2), $ROC(=S)NR_2$, where the carbonyl group (C=O) is replaced with a thiocarbonyl group (C=S)
- *S*-thiocarbamates (3), $RSC(=O)NR_2$, where the R–O– group is replaced with an R–S– group

O-thiocarbamates can isomerise to *S*-thiocarbamates, for example in the Newman-Kwart rearrangement.

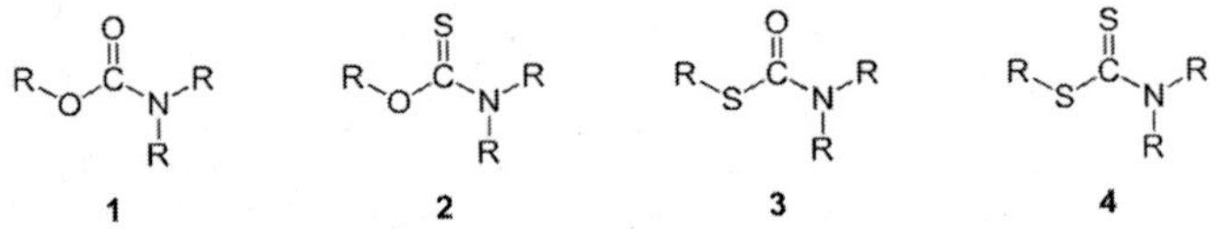

1 2 3 4

Carboxamide

In organic chemistry carboxamides (or amino carbonyls) are functional groups with the general structure $R\text{-}CO\text{-}NH_2$ with R as an organic substituent.

Two amino acids, asparagine and glutamine, have a carboxamide group in them. The properties and reactivity of the carboxamide group arise from the hydrogen bonding capabilities of the -NH2 group as well as the carbonyl oxygen. Furthermore, the carbon atom in a carboxamide has a low lying LUMO that is capable of accepting electron density from the nonbonding lone pair on the nitrogen, weakening the carbon-oxygen bond.

Dicarboximide Fungicides

Dicarboximide (or dicarboxamide) fungicides are a family of agricultural fungicides that include vinclozolin, iprodione, and procymidone. Dicarboximides are believed to inhibit triglyceride biosynthesis in sclerotia-forming fungi, including *Botrytis cinerea*. These fungicides turn into 3,5-dichloroaniline in soil rapidly. Repeated use of dicarboximides over several years reduce their effectiveness. Resistance has developed against all dicarboximides in many plant species, including vines, strawberries and protected crops, and are recommended to be used in conjunction with other fungicides.

Toxicity

Dicarboximides are endocrine disruptors and have been shown to have antiandrogenic effects, i.e. decrease levels of male hormones. Animal studies with vinclozolin and procymidone show irregular reproductive development due to their function as androgen receptor antagonists that inhibit androgen-activated gene expression. Even with low doses of antiandrogenic pesticides, developmental effects such as reduced anogenital distance and induction of areolas were seen in male rats.

Imidazole

Imidazole is an organic compound with the formula $(CH)_2N(NH)CH$. It is a colourless solid that dissolves in water to give mildly basic solution. In chemistry, it is an aromatic heterocycle, classified as a diazole and as an alkaloid. Derivatives of imidazole, called imidazoles, a common family heterocycles with sharing the 1,3-C3N2 ring, but varying substituents. This ring system is present in important biological building-blocks, such as histidine, and the related hormone histamine. Many drugs contain an imidazole ring, such as antifungal drugs, nitroimidazole, and the sedative midazolam.

Structure and Properties

Imidazole is a planar 5-membered ring. It exists in two equivalent tautomeric forms, because the proton can be located on either of the two nitrogen atoms. Imidazole is a highly polar compound, as evidenced by a calculated dipole of 3.61D. It is highly soluble in water. The compound is classified as aromatic due to the presence of a sextet of ∂-electrons, consisting of a pair of electrons from the protonated nitrogen atom and one from each of the remaining four atoms of the ring. Some resonance structures of imidazole are shown below:

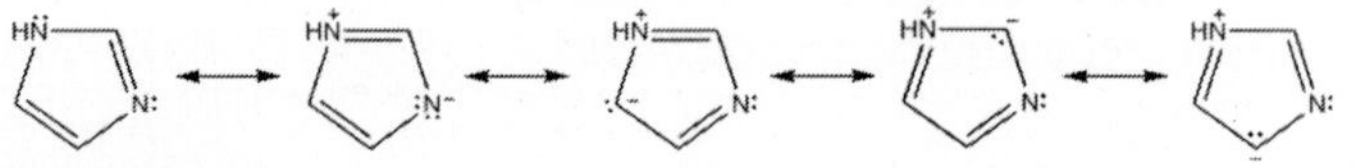

Amphotericity

Imidazole is amphoteric. That is, it can function as both an acid and as a base. As an acid, the pK_a of imidazole is 14.5, making it less acidic than carboxylic acids, phenols, and imides, but slightly more acidic than alcohols. The acidic proton is located on N-1.

As a base, the pK_a of the conjugate acid (cited above as p$K_{BH}{}^+$ to avoid confusion between the two) is approximately 7, making imidazole approximately sixty times more basic than pyridine. The basic site is N-3. Protonation gives the imidazolium cation, which is highly symmetric.

Preparation

Imidazole was first reported in 1858, although various imidazole derivatives had been discovered as early as the 1840s. Its synthesis, as shown below, used glyoxal and formaldehyde in ammonia to form imidazole (or glyoxaline, as it was originally named). This synthesis, while producing relatively low yields, is still used for creating C-substituted imidazoles.

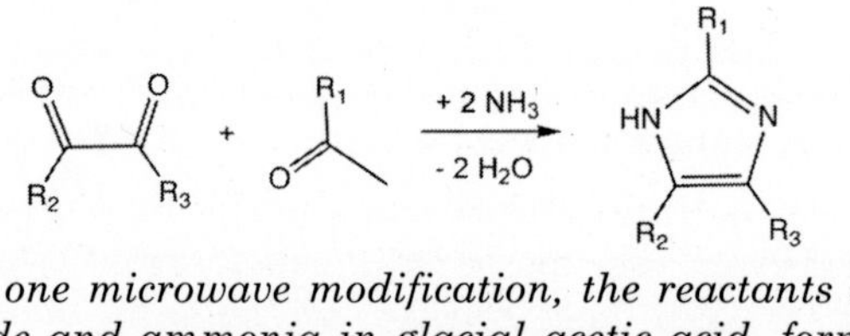

***Figure:** In one microwave modification, the reactants are benzil, benzaldehyde and ammonia in glacial acetic acid, forming 2,4,5-triphenylimidazole (Lophine).*

Imidazole can be synthesized by numerous methods besides the Debus method. Many of these syntheses can also be applied to different substituted imidazoles and imidazole derivatives by varying the functional groups on the reactants. These methods are commonly categorized by which and how many bonds form to make the imidazole rings. For example, the Debus method forms the (1,2), (3,4), and (1,5) bonds in imidazole, using each reactant as a fragment of the ring, and thus this method would be a three-bond-forming synthesis. A small sampling of these methods is presented below.

Formation of One Bond

The (1,5) or (3,4) bond can be formed by the reaction of an imidate and an á-aminoaldehyde or á-aminoacetal, resulting in the cyclization of an amidine to imidazole. The example below applies to imidazole when $R=R_1$=Hydrogen.

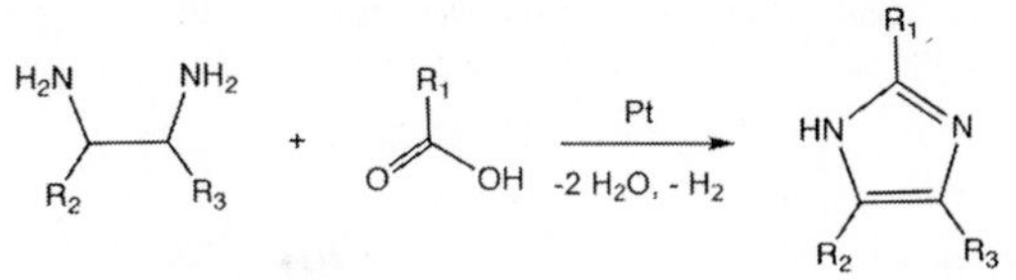

The (1,2) and (3,4) bonds can also be formed from N-substituted á-aminoketones and formamide with heat. The product will be a 1,4-disubstituted imidazole, but here since $R=R_1$=Hydrogen, imidazole itself is the product. The yield of this reaction is moderate, but it seems to be the most effective method of making the 1,4 substitution.

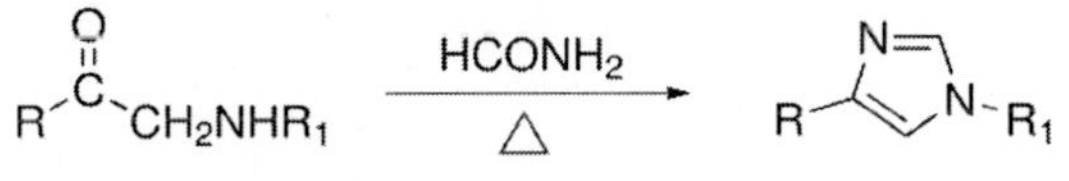

Formation of Four Bonds

This is a general method that is able to give good yields for substituted imidazoles.

In essence, it is an adaptation of the Debus method called the Debus-Radziszewski imidazole synthesis. The starting materials are substituted glyoxal, aldehyde, amine, and ammonia or an ammonium salt.

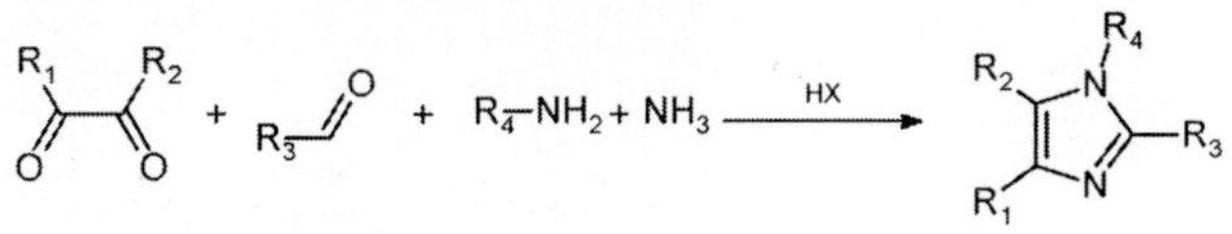

Formation from Other Heterocycles

Imidazole can be synthesized by the photolysis of 1-vinyltetrazole. This reaction will give substantial yields only if the 1-vinyltetrazole is made efficiently from an organotin compound, such as 2-tributylstannyltetrazole. The reaction, shown below, produces imidazole when $R=R_1=R_2$=Hydrogen.

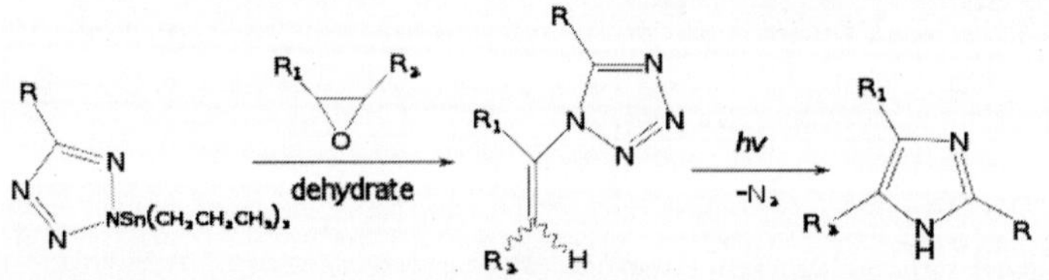

Imidazole can also be formed in a vapor-phase reaction. The reaction occurs with formamide, ethylenediamine, and hydrogen over platinum on alumina, and it must take place between 340 and 480°C. This forms a very pure imidazole product.

Chapter 2

Biological Significance and Applications

Imidazole is incorporated into many important biological molecules. The most pervasive is the amino acid histidine, which has an imidazole side-chain. Histidine is present in many proteins and enzymes and plays a vital part in the structure and binding functions of hemoglobin. Histidine can be decarboxylated to histamine, which is also a common biological compound. It is a component of the toxin that causes urticaria, which is another name for allergic hives. The relationship between histidine and histamine are shown below:

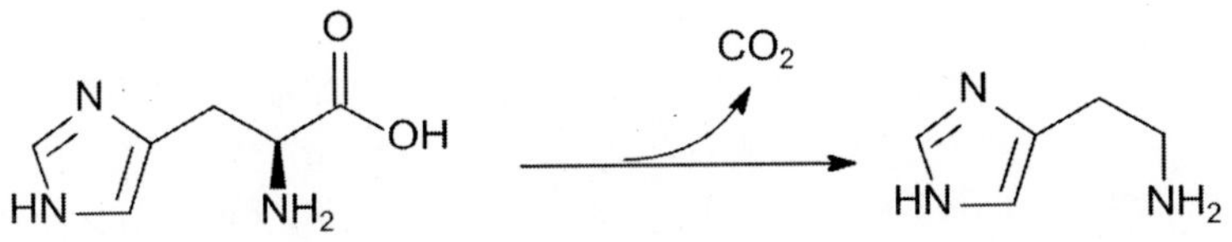

One of the applications of imidazole is in the purification of His-tagged proteins in immobilised metal affinity chromatography (IMAC). Imidazole is used to elute tagged proteins bound to Ni ions attached to the surface of beads in the chromatography column. An excess of imidazole is passed through the column, which displaces the His-tag from nickel co-ordination, freeing the His-tagged proteins.

Imidazole has become an important part of many pharmaceuticals. Synthetic imidazoles are present in many fungicides and antifungal, antiprotozoal, and antihypertensive medications. Imidazole is part of the theophylline molecule, found in tea leaves and coffee beans, that stimulates the central nervous system. It is present in the anticancer medication mercaptopurine, which combats leukemia by interfering with DNA activities.

Pharmaceutical Derivatives

The substituted imidazole derivatives are valuable in treatment of many systemic fungal infections. Imidazoles belong to the class of azole antifungals, which includes ketoconazole, miconazole, and clotrimazole. For comparison, another group of azoles is the triazoles, which includes fluconazole, itraconazole, and voriconazole. The difference between the imidazoles and the triazoles involves the mechanism of inhibition of the cytochrome P450 enzyme. The N3 of the imidazole compound binds to the heme iron atom of ferric cytochrome P450, whereas the N4 of the triazoles bind to the heme group. The triazoles have been shown to have a higher specificity for the cytochrome P450 than imidazoles, thereby making them more potent than the imidazoles.

Industrial Applications

Imidazole has been used extensively as a corrosion inhibitor on certain transition metals, such as copper. Preventing copper corrosion is important, especially in aqueous systems, where the conductivity of the copper decreases due to corrosion. Many compounds of industrial and technological importance contain imidazole derivatives. The thermostable polybenzimidazole PBI contains imidazole fused to a benzene ring and linked to a benzene, and acts as a fire retardant. Imidazole can also be found in various compounds that are used for photography and electronics.

Salts of Imidazole

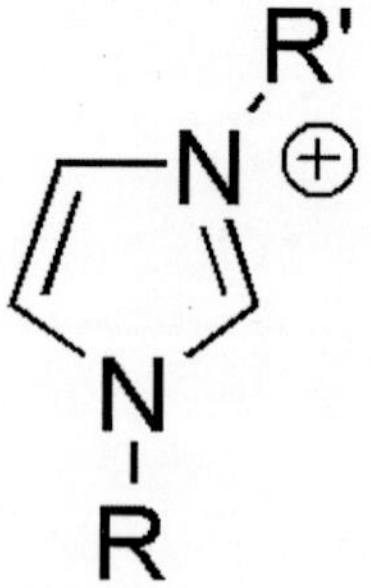

Salts of imidazole where the imidazole ring is in the cation are known as imidazolium salts (for example, imidazolium chloride). These salts are formed from the protonation or substitution at nitrogen of imidazole. These salts have been used as ionic liquids and precursors to stable carbenes. Salts where a deprotonated imidazole is an anion are also possible; these salts are known as imidazolide or imidazolate salts (for example, sodium imidazolide).

Related Heterocycles

- Benzimidazole, an analog with a fused benzene ring
- Dihydroimidazole or benzimidazoline, an analog where 4,5-double bond is saturated
- Pyrrole, an analog with only one nitrogen atom in position 1
- Oxazole, an analog with the nitrogen atom in position 1 replaced by oxygen
- Thiazole, an analog with the nitrogen atom in position 1 replaced by sulfur
- Pyrazole, an analog with two adjacent nitrogen atoms
- Triazoles, analogs with three nitrogen atoms

Isoxazole

Isoxazole is an azole with an oxygen atom next to the nitrogen. It is also the class of compounds containing this ring. Isoxazolyl is the univalent radical derived from isoxazole. Isoxazole rings are found in some natural products, such as ibotenic acid. Isoxazoles also form the basis for a number of drugs, including the COX-2 inhibitor valdecoxib (Bextra). A derivative, furoxan, is a nitric oxide donor. An isoxazolyl group is found in many beta-lactamase-resistant antibiotics, such as cloxacillin, dicloxacillin and flucloxacillin. The synthetic androgenic steroid danazol also has an isoxazole ring.

Morpholine

Morpholine is an organic chemical compound having the chemical formula $O(CH_2CH_2)_2NH$. This heterocycle, pictured at right, features both amine and ether functional groups. Because of the amine, morpholine is a base; its conjugate acid is called morpholinium. For example, neutralizing morpholine with hydrochloric acid makes the salt morpholinium chloride.

Production

Morpholine may be produced by the dehydration of diethanolamine with sulfuric acid:

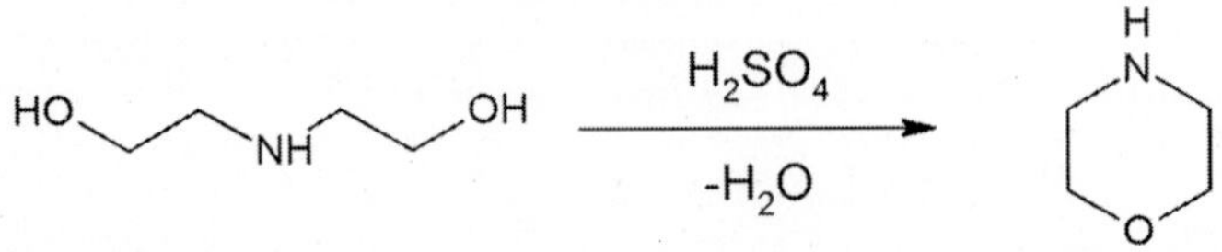

Uses

Industrial Applications: Morpholine is a common additive, in parts per million concentrations, for pH adjustment in both fossil fuel and nuclear power plant steam systems. Morpholine is used because its volatility is about the same as water, so once it is added to the water, its concentration becomes distributed rather evenly in both the water and steam phases.

Its pH adjusting qualities then become distributed throughout the steam plant to provide corrosion protection. Morpholine is often used in conjunction with low concentrations of hydrazine or ammonia to provide a comprehensive all-volatile treatment chemistry for corrosion protection for the steam systems of such plants. Morpholine decomposes reasonably slowly in the absence of oxygen at the high temperatures and pressures in these steam systems.

Organic Synthesis

Morpholine undergoes most chemical reactions typical for other secondary amines, though the presence of the ether oxygen withdraws electron density from the nitrogen, rendering it less nucleophilic (and less basic) than structurally similar secondary amines such as piperidine. For this reason, it forms a stable chloramine (CAS#23328-69-0).

It is Commonly Used to Generate Enamines

Morpholine is widely used in organic synthesis. For example, it is a building block in the preparation of the antibiotic linezolid, the anticancer agent gefitinib (Iressa) and the analgetic dextromoramide.

In research and in industry, the low cost and polarity of morpholine lead to its common use as a solvent for chemical reactions.

Producers

Most (as above) is widespread in Europe and the USA; accordingly, producers in Europe and the USA are able to cover the domestic and export markets.

Agriculture

As a Fruit Coating: Morpholine is used as a chemical emulsifier in the process of waxing fruit. Naturally, fruits make waxes to protect against insects and fungal contamination, but this can be lost as the fruit is cleaned. A small amount of new wax is applied to replace it. Morpholine is used as an emulsifier and solubility aid for shellac, which is used as a wax for fruit coating.

The European Union has forbidden the use of morpholine in fruit coating.

As a Component in Fungicides

Morpholine derivatives used as agricultural fungicides in cereals are known as ergosterol biosynthesis inhibitors.

- Amorolfine
- Fenpropimorph
- Tridemorph

Organophosphate

An organophosphate (sometimes abbreviated OP) is the general name for esters of phosphoric acid. Phosphates are probably the most pervasive organophosphorus compounds. Many of the most important biochemicals are organophosphates, including DNA and RNA as well as many cofactors that are essential for life. Organophosphates are the basis of many insecticides, herbicides, and nerve gases. The EPA lists organophosphates as very highly acutely toxic to bees, wildlife, and humans. Recent studies suggest a possible link to adverse effects in the neurobehavioural development of fetuses and children, even at very low levels of exposure. Organophosphates are widely used as solvents, plasticizers, and EP additives.

Organophosphates are widely employed both in natural and synthetic applications because of the ease with which organic groups can be linked together. Being a triprotic acid, phosphoric acid can form triesters whereas carboxylic acids only form monoesters. Esterification entails the attachment of organic groups to phosphorus through oxygen linkers. The precursors to such esters are alcohols. Encompassing many thousands of natural and synthetic compounds, alcohols are diverse and widespread.

$$OP(OH)_3 + ROH \rightarrow OP(OH)_2(OR) + H_2O$$

$$OP(OH)_2(OR) + R'OH \rightarrow OP(OH)(OR)(OR') + H_2O$$

$$OP(OH)(OR)(OR') + R''OH \rightarrow OP(OR)(OR')(OR'') + H_2O$$

The phosphate esters bearing OH groups are acidic and partially deprotonated in aqueous solution. For example, DNA and RNA are polymers of the type $[PO_2(OR)(OR')^-]_n$. Polyphosphates also form esters; an important example of an ester of a polyphosphate is ATP, which is the monoester of triphosphoric acid ($H_5P_3O_{10}$).

Alcohols can be detached from phosphate esters by hydrolysis, which is the reverse of the above reactions. For this reason, phosphate esters are common carriers of organic groups in biosynthesis.

Organophosphate Pesticides

In health, agriculture, and government, the word "organophosphates" refers to a group of insecticides or nerve agents acting on the enzyme acetylcholinesterase (the pesticide group carbamates also act on this enzyme, but through a different mechanism). The term is used often to describe virtually any organic phosphorus(V)-containing compound, especially when dealing with neurotoxic compounds. Many of the so-called organophosphates contain C-P bonds. For instance, sarin is *O*-isopropyl methylphosphonofluoridate, which is formally derived from phosphorous acid ($HP(O)(OH)_2$), not phosphoric acid ($P(O)(OH)_3$). Also, many compounds which are derivatives of phosphinic acid are used as neurotoxic organophosphates.

Organophosphate pesticides (as well as sarin and VX nerve agent) irreversibly inactivate acetylcholinesterase, which is essential to nerve function in insects, humans, and many other animals. Organophosphate pesticides affect this enzyme in varied ways, and thus in their potential for poisoning. For instance, parathion, one of the first OPs commercialized, is many times more potent than malathion, an insecticide used in combatting the Mediterranean fruit fly (Med-fly) and West Nile Virus-transmitting mosquitoes.

Organophosphate pesticides degrade rapidly by hydrolysis on exposure to sunlight, air, and soil, although small amounts can be detected in food and drinking water. Their ability to degrade made them an attractive alternative to the persistent organochloride pesticides, such as DDT, aldrin, and dieldrin. Although organophosphates degrade faster than the organochlorides, they have greater acute toxicity, posing risks to people who may be exposed to large amounts.

Commonly used organophosphates have included parathion, malathion, methyl parathion, chlorpyrifos, diazinon, dichlorvos, phosmet, fenitrothion, tetrachlorvinphos, azamethiphos, and azinphos methyl. Malathion is widely used in agriculture, residential landscaping, public recreation areas, and in public health pest control programs such as mosquito eradication. In the US, it is the most commonly used organophosphate insecticide. Forty organophosphate pesticides are registered in the U.S., with at least 73 million pounds used in agricultural and residential settings.

They are of concern to both scientists and regulators because they work by irreversibly blocking an enzyme that's critical to nerve function in both bugs and humans. Even at relatively low levels, organophosphates may be most hazardous to the brain development of fetuses and young children. The EPA banned most residential uses of organophosphates in 2001, but they are still sprayed agriculturally on fruits and vegetables. They're also used to control pests like mosquitos in public spaces such as parks. They can be absorbed through the lungs or skin or by eating them on food.

Organophosphates as Nerve Agents

History of Nerve Agents: Early pioneers in the field include Jean Louis Lassaigne (early 19th century) and Philippe de Clermont (1854). In 1932, German chemist Willy Lange and his graduate student, Gerde von Krueger, first described the cholinergic nervous system effects of organophosphates, noting a choking sensation and a dimming of vision after exposure. This discovery later inspired German chemist Gerhard Schrader at company IG Farben in the 1930s to experiment with these compounds as insecticides. Their potential use as chemical warfare agents soon became apparent, and the Nazi government put Schrader in charge of developing organophosphate (in the broader sense of the word) nerve gases.

Schrader's laboratory discovered the G series of weapons, which included Sarin, Tabun, and Soman. The Nazis produced large quantities of these compounds, though did not use them during World War II. British scientists experimented with a cholinergic organophosphate of their own, called diisopropylfluorophosphate (DFP), during the war. The British later produced VX nerve agent, which was many times more potent than the G series, in the early 1950s, almost 20 years after the Germans had discovered the G series. After World War II, American companies gained access to some information from Schrader's laboratory, and began synthesizing organophosphate pesticides in large quantities. Parathion was among the first marketed, followed by malathion and azinphosmethyl. The popularity of these insec-ticides increased after many of the organochlorine insecticides like DDT, dieldrin, and heptachlor were banned in the 1970s.

Structural Features of Organophosphates

Effective organophosphates have the following structural features:

- A terminal oxygen connected to phosphorus by a double bond, i.e. a phosphoryl group
- Two lipophilic groups bonded to the phosphorus
- A leaving group bonded to the phosphorus, often a halide

Terminal Oxygen vs. Terminal Sulfur

Thiophosphoryl compounds, those bearing the P=S functionality, are much less toxic than related phosphoryl derivatives, which include sarin, VX and tetraethyl pyrophosphate. Thiophosphoryl compounds are not active inhibitors of acetylcholinesterase in either mammals or insects; in mammals, metabolism tends to remove lipophilic side groups from the phosphorus atom while in insects it tends to oxidize the compound, thus removing the terminal sulfur and replacing it with a terminal oxygen, which allows the compound to more efficiently act as an acetylcholinesterase inhibitor.

Fine Tuning

Within these requirements, a large number of different lipophilic and leaving groups have been used. The variation of these groups is one means of fine tuning the toxicity of the compound. A good example of this chemistry are the *P*-thiocyanate compounds which use an aryl (or alkyl) group and an alkylamino group as the lipophilic groups. The thiocyanate is the leaving group.

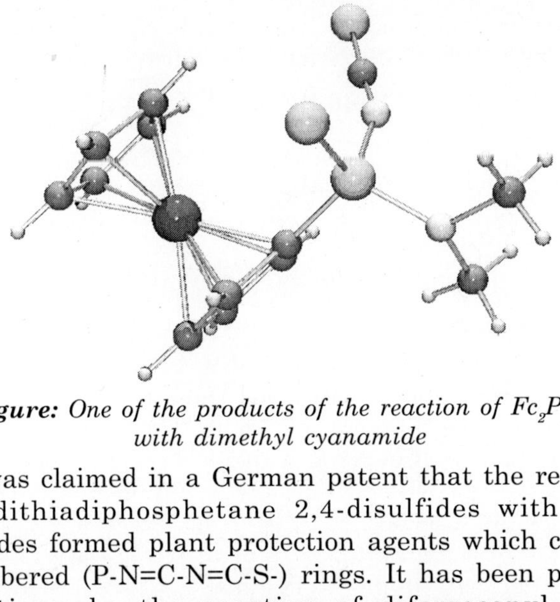

Figure: *One of the products of the reaction of* $Fc_2P_2S_4$ *with dimethyl cyanamide*

It was claimed in a German patent that the reaction of 1,3,2,4-dithiadiphosphetane 2,4-disulfides with dialkyl cyanamides formed plant protection agents which contained six membered (P-N=C-N=C-S-) rings. It has been proven in recent times by the reaction of diferrocenyl 1,3,2,4-dithiadiphosphetane 2,4-disulfide (and Lawesson's reagent) with dimethyl cyanamide that, in fact, a mixture of several different phosphorus-containing compounds is formed.

Depending on the concentration of the dimethyl cyanamide in the reaction mixture, either a different six membered ring compound (P-N=C-S-C=N-) or a nonheterocylic compound (FcP(S)(NR2)(NCS)) is formed as the major product; the other compound is formed as a minor product.

In addition, small traces of other compounds are also formed in the reaction. It is unlikely that the ring compound (P-N=C-

S-C=N-) {or its isomer} would act as a plant protection agent, but (FcP(S)(NR2)(NCS)) compounds can act as nerve poisons in insects.

Organophosphate Poisoning

Many *organophosphates* are potent nerve agents, functioning by inhibiting the action of acetylcholinesterase (AChE) in nerve cells. They are one of the most common causes of poisoning worldwide, and are frequently intentionally used in suicides in agricultural areas.

Organophosphorus pesticides can be absorbed by all routes, including inhalation, ingestion, and dermal absorption. Their toxicity is not limited to the acute phase, however, and chronic effects have long been noted. Neurotransmitters such as acetylcholine (which is affected by organophosphate pesticides) are profoundly important in the brain's development, and many OPs have neurotoxic effects on developing organisms, even from low levels of exposure. Other organophosphates are not toxic, yet their main metabolites, such as their oxons are.

Health Effects

Chronic Toxicity: Repeated or prolonged exposure to organophosphates may result in the same effects as acute exposure including the delayed symptoms. Other effects reported in workers repeatedly exposed include impaired memory and concentration, disorientation, severe depressions, irritability, confusion, headache, speech difficulties, delayed reaction times, nightmares, sleepwalking and drowsiness or insomnia. An influenza-like condition with headache, nausea, weakness, loss of appetite, and malaise has also been reported.

Low Level Exposure

Even at relatively low levels organophosphates may be hazardous to human health. The pesticides act on a set of brain chemicals closely related to those involved in ADHD, thus fetuses and young children, where brain development depends on a strict sequence of biological events, may be most at risk.

They can be absorbed through the lungs or skin or by eating them on food. According to a 2008 report from the U.S. Department of Agriculture, in a representative sample of produce tested by the agency, 28 percent of frozen blueberries,

20 percent of celery, 27 percent of green beans, 17 percent of peaches, 8 percent of broccoli and 25 percent of strawberries contained traces of organophosphate.

The United States Environmental Protection Agency lists the organophosphate parathion as a possible human carcinogen.

Chronic fatigue is common amongst those who consider their health is affected by pesticides and research from 2003 suggested there was an association between exposure to organophosphates and chronic fatigue symptoms.

A 2007 study linked the organophosphate insecticide chlorpyrifos, which is used on some fruits and vegetables, with delays in learning rates, reduced physical coordination, and behavioural problems in children, especially ADHD.

An organic diet is an effective way to reduce exposure to the organophosphorus pesticides that are commonly used in agricultural production. Organophosphate metabolite levels rapidly drop and, for some metabolites, become undetectable in children's urine when an organic diet is consumed.

A 2010 study has found that organophosphate exposure is associated with an increased risk of Alzheimer's disease. Another study from the same year found that each 10-fold increase in urinary concentration of organophosphate metabolites was associated with a 55% to 72% increase in the odds of ADHD in children.

The study found that organophosphate exposure is associated with an increased risk of ADHD in children. Researchers analyzed the levels of organophosphate residues in the urine of more than 1,100 children aged 8 to 15 years old, and found that those with the highest levels of dialkyl phosphates, which are the breakdown products of organophosphate pesticides, also had the highest incidence of ADHD.

Overall, they found a 35% increase in the odds of developing ADHD with every 10-fold increase in urinary concentration of the pesticide residues.

The effect was seen even at the low end of exposure: children who had any detectable, above-average level of pesticide metabolite in their urine were twice as likely as those with undetectable levels to record symptoms of ADHD.

Another 2010 study found that children who were exposed to organophosphate pesticides while still in their mother's womb were more likely to develop attention disorders years later. The researchers evaluated the children at age 3.5 and 5 years for symptoms of attention disorders and ADHD using maternal reports of child behaviour, performance on standardized computer tests, and behaviour ratings from examiners. Each tenfold increase in prenatal pesticide metabolites was linked to having five times the odds of scoring high on the computerized tests at age 5, suggesting a greater likelihood of a child having ADHD. The effect appeared to be stronger for boys than for girls.

A 2012 study published in *Environmental Health Perspectives* found that prenatal organophosphate exposure had a significant impact on birthweight and gestational age. A 10-fold increase in organophosphates concentrations in the mother was associated with a 0.5 week decrease in the infant's gestational age and a birthweight decline of 151 grams (adjusted to account for changes in gestational age). "Diet and home pesticide use have been identified as important routes of exposure in non-agricultural populations," the researchers wrote, but noted that switching children from conventional to organic diets for several days reduced levels near or below the limit of detection, "suggesting that diet was the primary source of exposure in that study population."

Proposals to Ban and Restrictions of Use

According to the non-governmental organisation Pesticide Action Network, the organophosphate parathion is one of the most dangerous pesticides. In the US alone more than 650 agricultural workers have been poisoned since 1966, of which 100 died. In underdeveloped countries many more people have suffered fatal and nonfatal intoxications. The World Health Organization, PAN and numerous environmental organisations propose a general and global ban. Its use is banned or restricted in 23 countries and its import is illegal in a total of 50 countries. Its use was banned in the U.S. in 2000 and it has not been used since 2003.

Other than for agricultural use, the organophosphate diazinon has been banned in the U.S. Agriculturally, more than one million pounds of diazinon were used in California to control agricultural pests in 2000. The areas and crops on which diazinon are most heavily applied are structural pest control, almonds, head lettuce, leaf lettuce and prunes.

In May 2006 the Environmental Protection Agency reviewed the use of the organophosphate dichlorvos and proposed its continued sale, despite concerns over its safety and considerable evidence suggesting it is carcinogenic and harmful to the brain and nervous system, especially in children. Environmentalists charge that the latest decision was the product of backroom deals with industry and political interference.

In 2001 the EPA placed new restrictions on the use of the organophosphates phosmet and azinphos-methyl to increase protection of agricultural workers. The crop uses reported at that time as being phased out in four years included those for almonds, tart cherries, cotton, cranberries, peaches, pistachios, and walnuts. The crops with time-limited registration included apples/crab apples, blueberries, sweet cherries, pears, pine seed orchards, brussels sprouts, cane berries, and the use of azinphos-methyl by nurseries for quarantine requirements. The labelled uses of phosmet include alfalfa, orchard crops (e.g. almonds, walnuts, apples, cherries), blueberries, citrus, grapes, ornamental trees (not for use in residential, park, or recreational areas) and non-bearing fruit trees, Christmas trees and conifers (tree farms), potatoes and peas. Azinphos-methyl has been banned in Europe since 2006.

Aniline

Aniline, phenylamine or aminobenzene is an organic compound with the formula $C_6H_5NH_2$. Consisting of a phenyl group attached to an amino group, aniline is the prototypical aromatic amine. Being a precursor to many industrial chemicals, its main use is in the manufacture of precursors to polyurethane. Like most volatile amines, it possesses the somewhat unpleasant odor of rotten fish. It ignites readily, burning with a smoky flame characteristic of aromatic compounds. Aniline is colourless, but it slowly oxidizes and resinifies in air, giving a red-brown tint to aged samples.

Production

Aniline is mainly produced in industry in two steps from benzene. First, benzene is nitrated using a concentrated mixture of nitric acid and sulfuric acid at 50 to 60°C, which gives nitrobenzene. In the second step, the nitrobenzene is hydrogenated, typically at 200-300 °C in presence of various metal catalysts:

$$C_6H_5NO_2 + 3\ H_2 \rightarrow C_6H_5NH_2 + 2\ H_2O$$

Originally, the reduction was effected with a mixture of ferrous chloride and iron metal via the Bechamp reduction.

As an alternative, aniline is also prepared from phenol and ammonia, the phenol being derived from the cumene process.

In commerce, three brands of aniline are distinguished: aniline oil for blue, which is pure aniline; aniline oil for red, a mixture of equimolecular quantities of aniline and ortho- and para-toluidines; and aniline oil for safranine, which contains aniline and ortho-toluidine, and is obtained from the distillate (échappés) of the fuchsine fusion.

Related Aniline Derivatives

Many derivatives of aniline can be prepared in similar fashion from nitrated aromatic compounds. Nitration followed by reduction of toluene affords toluidines. Nitration of chlorobenzene and related derivatives and reduction of the nitration products gives aniline derivatives, e.g. 4-chloroaniline.

Reactions

The chemistry of aniline is extremely rich because the compound has been cheaply available for many years. Below are some classes of its reactions.

Oxidation

The oxidation of aniline has been heavily investigated, and can result in reactions localized at nitrogen or more commonly results in the formation of new C-N bonds. In alkaline solution, azobenzene results, whereas arsenic acid produces the violet-colouring matter violaniline. Chromic acid converts it into quinone, whereas chlorates, in the presence of certain metallic salts (especially of vanadium), give aniline black. Hydrochloric acid and potassium chlorate give chloranil. Potassium permanganate in neutral solution oxidizes it to nitrobenzene, in alkaline solution to azobenzene, ammonia and oxalic acid, in acid solution to aniline black. Hypochlorous acid gives 4-

aminophenol and para-amino diphenylamine. Oxidation with persulfate affords a variety of polyanilines compounds. These polymers exhibit rich redox and acid-base properties.

Electrophilic Reactions at Carbon

Like phenols, aniline derivatives are highly susceptible to electrophilic substitution reactions. Its high reactivity reflects that it is an enamine, which enhances the electron-donating ability of the ring. For example, reaction of aniline with sulfuric acid at 180 °C produces sulfanilic acid, $H_2NC_6H_4SO_3H$, which can be converted to sulfanilamide. Sulfanilamide is one of the sulfa drugs, which were widely used as antibacterials in the early 20th century. The largest scale industrial reaction of aniline involves its alkylation with formaldehyde:

$$2\ C_6H_5NH_2 + CH_2O \rightarrow CH_2(C_6H_4NH_2)_2 + H_2O$$

The resulting diamine is the precursor to 4,4'-MDI and related diisocyanates.

Reactions at Nitrogen

Basicity: Aniline is a weak base. Aromatic amines such as aniline are, in general, much weaker bases than aliphatic amines because of the electron-withdrawing effect of the phenyl group. Aniline reacts with strong acids to form anilinium (or phenylammonium) ion (C_6H_5-NH_3^+). Although aniline is weakly basic, it precipitates zinc, aluminium, and ferric salts, and, on warming, expels ammonia from its salts. The weak basicity is due to a negative inductive effect as the lone pair on the nitrogen is partially delocalized into the pi system of the benzene ring.

Acylation

Aniline reacts with carboxylic acids or more readily with acyl chlorides such as acetyl chloride to give amides. The amides formed from aniline are sometimes called anilides, for example CH_3-CO-NH-C_6H_5 is acetanilide. Antifebrin (acetanilide), an anti-pyretic and analgesic, is obtained by the reaction of acetic acid and aniline.

N-Alkylation

N-Methylation of aniline with methanol at elevated temperatures over acid catalysts gives *N*-methylaniline and dimethylaniline:

$$C_6H_5NH_2 + 2\ CH_3OH \rightarrow C_6H_5N(CH_3)_2 + 2H_2O$$

N-Methylaniline and dimethylaniline are colourless liquids with boiling points of 193-195 °C and 192 °C, respectively. These derivatives are of importance in the colour industry. Aniline combines directly with alkyl iodides to form secondary and tertiary amines.

Carbon Disulfide Derivatives

Boiled with carbon disulfide, it gives sulfocarbanilide (diphenylthiourea) ($CS(NHC_6H_5)_2$), which may be decomposed into phenyl isothiocyanate(C_6H_5CNS), and triphenyl guanidine ($C_6H_5N{=}C(NHC_6H_5)_2$).

Diazotization

Aniline and its ring-substituted derivatives react with nitrous acid to form diazonium salts. Through these intermediates, aniline can be conveniently converted to -OH, -CN, or a halide via Sandmeyer reactions. This diazonium salt can also be reacted with NaNO2 and phenol which produces a dye which is benzeneazophenol, this process is called coupling.

Other Reactions

It reacts with nitrobenzene to produce phenazine in the Wohl-Aue reaction. Hydrogenation gives cyclohexylamine.

Being a standard reagent in laboratories, aniline is used for many niche reactions. Its acetate is used in the Aniline acetate test for carbohydrates, identifying pentoses by conversion to furfural. It is used to stain neural RNA blue in the Nissl stain.

Uses

The largest application of aniline is for the preparation of methylene diphenyl diisocyanate (MDI). The majority of aniline serves this market. Other uses include rubber processing chemicals (9%), herbicides (2%), and dyes and pigments (2%). As additives to rubber, aniline derivatives such as phenylene-diamines and diphenylamine, are antioxidants. Illustrative of the drugs prepared from aniline is paracetamol (acetaminophen, Tylenol). The principal use of aniline in the dye industry is as a precursor to indigo, the blue of blue jeans.

Aniline is also used at a smaller scale in the production of the intrinsically conducting polymer polyaniline.

History

Aniline was first isolated by destructive distillation of indigo by Otto Unverdorben, who named it *Crystallin*. In 1834, Friedlieb Runge isolated from coal tar a substance that turned a beautiful blue colour when treated with chloride of lime, and he named it *kyanol* or *cyanol*. In 1840, Carl Julius Fritzsche (1808-1871) treated indigo with caustic potash and obtained an oil that he named *aniline*, after an indigo-yielding plant, Añil (*Indigofera suffruticosa*). In 1842, Nikolay Nikolaevich Zinin reduced nitrobenzene and obtained a base that he named *benzidam*. In 1843, August Wilhelm von Hofmann showed that all of these substances are the same substance — thereafter known as *phenylamine* or *aniline*.

Synthetic Dye Industry

In 1856, von Hofmann's student William Henry Perkin discovered mauveine and went into industry producing the first synthetic dye. Other aniline dyes followed, such as fuchsine, safranine, and induline. At the time of mauveine's discovery, aniline was expensive. Soon thereafter, applying a method reported in 1854 by Antoine Béchamp, it was prepared "by the ton". The Béchamp reduction enabled the evolution of a massive dye industry in Germany. Today, the name of BASF, originally *Badische Anilin- und Soda-Fabrik*, now among the largest chemical suppliers, echoes the legacy of the synthetic dye industry, built via aniline dyes and extended via the related azo dyes. The first azo dye was aniline yellow.

Developments in Medicine

In the late 19th century, aniline emerged as an analgesic drug, its cardiac-suppressive side effects countered with caffeine. During the first decade of the 20th century, while trying to modify synthetic dyes to treat African sleeping sickness, Paul Ehrlich – who had coined the term *chemotherapy* for his *magic bullet* approach to medicine – failed and switched to modifying Béchamp's atoxyl, the first organic arsenical drug, and serendipitously obtained a treatement for syphilis – salvarsan – the first successful chemotherapy agent. Salvarsan's targeted microorganism, not yet recognized as a bacterium, was still thought to be a parasite, and medical bacteriologists, believing that bacteria were not susceptible to the chemotherapeutic approach, overlooked Alexander Fleming's report in 1928 on the effects of penicillin.

In 1932, Bayer sought medical applications of its dyes. Gerhard Domagk identified as an antibacterial a red azo dye, introduced in 1935 as the first antibacterial drug, prontosil, soon found at Pasteur Institute to be a prodrug degraded *in vivo* into sulfanilamide – a colourless intermediate for many, highly colourfast azo dyes – already with an expired patent, synthesized in 1908 in Vienna by the researcher Paul Gelmo for his doctoral research.

By the 1940s, over 500 related sulfa drugs were produced. Medications in high demand during World War II (1939 – 45), these first *miracle drugs*, chemotherapy of wide effectiveness, propelled the American pharmaceutics industry. In 1939, at Oxford University, seeking an alternative to sulfa drugs, Howard Florey developed Fleming's penicillin into the first systemic antibiotic drug, penicillin G. (Gramicidin, developed by René Dubos at Rockefeller Institute in 1939, was the first antibiotic, yet its toxicity restricted it to topical use.) After World War II, Cornelius P. Rhoads introduced the chemotherapeutic approach to cancer treatment.

Toxicology

Aniline is toxic by inhalation of the vapour, ingestion, or percutaneous absorption. The IARC lists it in Group 3 (*not classifiable as to its carcinogenicity to humans*) due to the limited and contradictory data available. The early manufacture of aniline resulted in increased incidents of bladder cancer, but these effects are now attributed to naphthylamines, not anilines.

Detection

Qualitative Test: It is metabolized to p-aminophenol and p-acetamidophenol, which are excreted in urine as sulfate and glucuronide conjugates. On hydrolysis of urine, p-aminophenol is reformed, and can be detected using the o-cresol test.

- Reagents
  1. Concentrated hydrochloric acid with relative density 1.18.
  2. Aqueous 10 g/l o-cresol.
  3. Aqueous 4 mol/l ammonium hydroxide.

- Method
  1. Add 0.5 ml of hydrochloric acid to 0.5 ml of sample, boil for 10 minutes and cool.
  2. Add 1 ml of o-cresol solution to 0.2 ml of the hydrolysate.
  3. Add 2 ml of ammonium hydroxide solution and mix for 5 seconds.
- Results

A strong, royal blue colour developing rapidly indicates the presence of p-aminophenol.

Metabolites of paracetamol and nitrobenzene also give p-aminophenol on hydrolysis and so interfere. Ethylenediamine gives a green colour in this test.

- Sensitivity

p-Aminophenol, 10 mg/l.

Confirmatory Test

Aniline and other primary aromatic amines form diazo compounds with nitrous acid, which couple with 1-naphthylethylenediamine to form highly coloured derivatives. This reaction forms the basis of the confirmatory test.

- Reagents
  1. Aqueous 2 g/l, freshly prepared sodium nitrite solution.
  2. Aqueous 2 mol/lhydrochloric acid.
  3. Aqueous 10 g/l ammonium sulfamate solution.
  4. Aqueous 2 g/l, freshly prepared N-(1-naphthyl) ethylenediamine dihydrochloride solution.
- Method
  1. Mix 0.1 ml of sodium nitrite solution and 0.2 ml of dilute hydrochloric acid in a 5 ml test-tube.
  2. Add 0.1 ml of sample, mix and allow to stand for 2 minutes.
  3. Add 0.2 ml of ammonium sulfamate solution followed by 0.1 ml of N-(1-naphthyl)ethylenediamine dihydrochloride solution.

- Results

A purple colour after 1 minute indicates the presence of aniline.

- Sensitivity

Aniline, 10 mg/l.

Phosphonate

Phosphonates or phosphonic acids are organic compounds containing C-PO$(OH)_2$ or C-PO$(OR)_2$ groups (where R=alkyl, aryl). Bisphosphonates were first synthesized in 1897 by Von Baeyer and Hofmann.

An example of such a bisphosphonate is HEDP (Etidronic acid or Didronel). Since the work of Schwarzenbach in 1949, phosphonic acids are known as effective chelating agents. The introduction of an amine group into the molecule to obtain -NH_2-C-PO$(OH)_2$ increases the metal binding abilities of the phosphonate.

Examples for such compounds are NTMP, EDTMP and DTPMP. These common phosphonates are the structure analogues to the well-known aminopolycarboxylates NTA, EDTA, and DTPA.

The stability of the metal complexes increases with increasing number of phosphonic acid groups. Phosphonates are highly water-soluble while the phosphonic acids are only sparingly soluble. Phosphonates are not volatile and are poorly soluble in organic solvents.

Chemistry

Phosphonates can be synthesized using the Michaelis–Arbuzov reaction. In one study a á-aminophosphonate is prepared by condensation of benzaldehyde, aniline, and trimethyl phosphite catalyzed by Copper triflate in a one-pot synthesis.

The phosphonates (and other phosphorus esters) can also be synthesized via transesterification reaction catalyzed by organic catalysts. N-heterocyclic carbenes have been reported to catalyze the reaction efficiently.

In organic synthesis, phosphonates are used in the Horner-Wadsworth-Emmons reaction.

Phosphonates are effective chelating agents that bind tightly to di- and trivalent metal ions, preventing them from forming insoluble precipitates (scale) and suppressing their catalytic properties. They are stable under harsh conditions.

An important industrial use of phosphonates is in cooling waters, desalination systems, and in oil fields to inhibit scale formation. Phosphonates are also regularly used in reverse osmosis systems as anti-scalants.

Phosphonates in cooling water systems also serve to control corrosion of iron and steel. In pulp and paper manufacturing and in textile industry they serve as "peroxide bleach stabilizers," by chelating metals that could inactivate the peroxide. In detergents they are used as a combination of chelating agent, scale inhibitor, and bleach stabilizer.

Phosphonates are also increasingly used in medicine to treat disorders associated with bone formation and calcium metabolism. Furthermore they serve as carriers for radionuclides in bone cancer treatments.

Occurrence in Nature

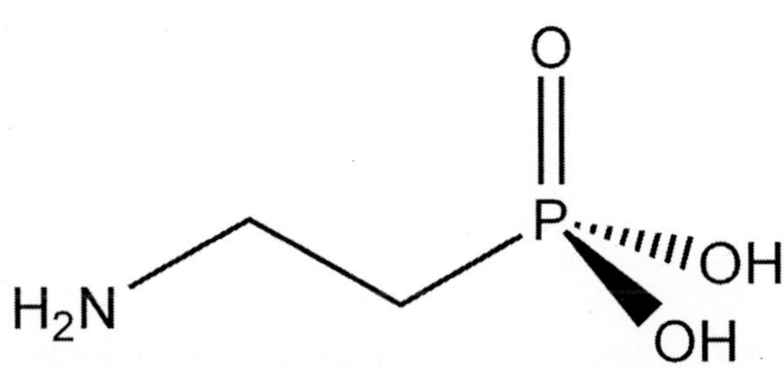

Figure: *2-aminoethylphosphonic acid: the first identified natural phosphonate.*

The naturally-occurring phosphonate 2-aminoethyl-phosphonic acid was first identified in 1959 in plants and many animals, where it is localized in membranes.

Phosphonates are quite common among different organisms, from prokaryotes to eubacteria and fungi, mollusks, insects and others. They were first reported in natural soils by Newman and Tate (1980). The biological role of the natural phosphonates is still poorly understood. Bis- or polyphosphonates have not been found to occur naturally.

Uses

In 1998 the consumption of phosphonates was 56,000 tons worldwide - 40,000 tons in the US, 15,000 tons in Europe and less than 800 tons in Japan. The demand of phosphonates grows steadily at 3% annually. In conjunction with organosilicates, phosphonates are also used to treat "sudden oak death", which is caused by the fungus-like eukaryote *Phytophthora ramorum*.

Toxicology

The toxicity of phosphonates to aquatic organisms is low. Reported values for 48 h LC50 values for fish are between 0.1 and 1.1 mM. Also the bioconcentration factor for fish is very low.

Biodegradation

In nature bacteria play a major role in the degradation of phosphonates. Due to the presence of natural phosphonates in the environment, bacteria have evolved the ability to metabolize phosphonates as nutrient sources. Some bacteria use phosphonates as a phosphorus source for growth. Aminophosphonates can also be used as sole nitrogen source by some bacteria.

The polyphosphonates used in industry differ greatly from natural phosphonates such as 2-aminoethylphosphonic acid, because they are much larger, carry a high negative charge and are complexed with metals. Biodegradation tests with sludge from municipal sewage treatment plants with HEDP and NTMP showed no indication for any degradation. An investigation of HEDP, NTMP, EDTMP and DTPMP in standard biodegradation tests also failed to identify any biodegradation.

It was noted, however, that in some tests due to the high sludge to phosphonate ratio, removal of the test substance from solution observed as loss of DOC was observed. This factor was attributed to adsorption rather than biodegradation. However, bacterial strains capable of degrading aminopolyphosphonates and HEDP under P-limited conditions have been isolated from soils, lakes, wastewater, activated sludge and compost. No biodegradation of phosphonates during water treatment is observed but photodegradation of the Fe(III)-complexes is rapid. Aminopolyphosphonates are also rapidly oxidized in the presence of Mn(II) and oxygen and stable breakdown products are formed that have been detected in wastewater. The lack of information about phosphonates in the environment is linked to analytical problems of their determi-nation at trace concentrations in natural waters. Phosphonates are present mainly as Ca and Mg-complexes in natural waters and therefore do not affect metal speciation or transport. Phosphonates are one of the three sources of phosphate intake in biological cells (The other two being inorganic phosphate and organophosphate)

Environmental Behaviour

Phosphonates have a very strong interaction with surfaces, which results in a significant removal in technical and natural systems. Due to this strong adsorption, little or no remobilization of metals is expected.

Phosphonate Compounds

- AEPn: 2-Aminoethylphosphonic acid
- DMMP: Dimethyl methylphosphonate
- HEDP: 1-Hydroxy Ethylidene-1,1-Diphosphonic Acid
- ATMP: Amino tris(methylene phosphonic acid)
- EDTMP: Ethylenediamine tetra(methylene phosphonic acid)
- TDTMP: Tetramethylenediamine tetra(methylene phosphonic acid)
- HDTMP: Hexamethylenediamine tetra(methylene phosphonic acid)
- DTPMP: Diethylenetriamine penta(methylene phosphonic acid)
- PBTC: Phosphonobutane-tricarboxylic acid
- PMIDA: N-(phosphonomethyl)iminodiacetic acid
- CEPA: 2-carboxyethyl phosphonic acid
- HPAA: 2-Hydroxyphosphonocarboxylic acid
- AMP: Amino-tris-(methylene-phosphonic acid)

Piperazine

Piperazine is an organic compound that consists of a six-membered ring containing two nitrogen atoms at opposite positions in the ring. Piperazine exists as small alkaline deliquescent crystals with a saline taste. The piperazines are a broad class of chemical compounds, many with important pharmacological properties, which contain a core piperazine functional group.

Origin and Naming

Piperazines were originally named because of their chemical similarity with piperidine, part of the structure of piperine in the black pepper plant (*Piper nigrum*). It is important to note, however, that piperazines are *not* derived from plants in the *Piper* genus.

Chemistry

Piperazine is freely soluble in water and ethylene glycol, but insoluble in diethyl ether. It is a weak base with a pK_b of 4.19; the pH of a 10% aqueous solution of piperazine is 10.8-11.8. Piperazine readily absorbs water and carbon dioxide from the air. Although many piperazine derivatives occur naturally, piperazine itself can be synthesized by reacting alcoholic ammonia with 1,2-dichloroethane, by the action of sodium and ethylene glycol on ethylene diamine hydrochloride, or by reduction of pyrazine with sodium in ethanol. A form in which piperazine is commonly available industrially is as the hexahydrate, $C_4H_{10}N_2$. $6H_2O$, which melts at 44°C and boils at 125-130°C. Two common salts in the form of which piperazine is usually prepared for pharmaceutical or veterinary purposes are the citrate, $3C_4H_{10}N_2.2C_6H_8O_7$ (i.e. containing 3 molecules of piperazine to 2 molecules of citric acid), and the adipate, $C_4H_{10}N_2.C_6H_{10}O_4$ (containing 1 molecule each of piperazine and adipic acid).

Industrial Production

Piperazine is formed as a co-product in the ammoniation of 1,2-dichloroethane or ethanolamine. These are the only routes to the chemical used commercially. The piperazine is separated from the product stream, which contains ethylenediamine, diethy-lenetriamine, and other related linear and cyclic chemicals of this type.

As an Anti-Helmintic

Piperazine was first introduced as an anthelmintic in 1953. A large number of piperazine compounds have anthelmintic action.

Their mode of action is generally by paralysing parasites, which allows the host body to easily remove or expel the invading organism.

This action is mediated by its agonist effects upon the inhibitory GABA (ã-aminobutyric acid) receptor. Its selectivity for helminths is because vertebrates only use GABA in the CNS and the helminths' GABA receptor is a different isoform to the vertebrate's one.

Piperazine hydrate and *piperazine citrate* are the main anthel-mintic piperazines. These drugs are often referred to simply as "piperazine" which may cause confusion between the specific anthel-mintic drugs and the entire class of piperazine-containing compounds.

Other Uses

Piperazines are also used in the manufacture of plastics, resins, pesticides, brake fluid and other industrial materials. Piperazines, especially BZP and TFMPP were extremely common adulterants in the club and rave scene, often being passed off as MDMA, although they do not share many similarities in their effects. Piperazine is also a fluid used for CO_2 and H_2S scrubbing in association with methyl diethanolamine (MDEA). In Pakistan, SB CHICK COMPANY Pvt.Ltd (Rawalpindi) have a product named SB PIPERAZINE as a poultry dewormer

Piperazine Derivatives as Drugs

Many currently notable drugs contain a piperazine ring as part of their molecular structure. Examples include:

Antianginals

- Ranolazine
- Trimetazidine

Antidepressants

- Amoxapine
- Befuraline
- Buspirone
- Flesinoxan
- Gepirone
- Ipsapirone

- Nefazodone
- Piberaline
- Tandospirone
- Trazodone
- Vilazodone
- Zalospirone

Antihistamines

- Buclizine
- Meclozine
- Cinnarizine
- Cyclizine
- Hydroxyzine
- Cetirizine
- Levocetirizine
- Niaprazine

Antipsychotics

- Typical
  * Fluphenazine
  * Perphenazine
  * Trifluoperazine
  * Prochlorperazine
  * Thiothixene
  * Flupentixol
  * Zuclopenthixol
- Atypical
  * Amperozide
  * Aripiprazole
  * Lurasidone
  * Clozapine
  * Olanzapine
  * Perospirone
  * Ziprasidone

Recreational Drugs

- 4-Bromo-2,5-dimethoxy-1-benzylpiperazine (2C-B-BZP)
- 1-Benzylpiperazine (BZP)
- 2,3-Dichlorophenylpiperazine (DCPP)
- 1,4-Dibenzylpiperazine (DBZP)
- 4-Methyl-1-benzylpiperazine (MBZP)
- 3-Chlorophenylpiperazine (mCPP)
- 3,4-Methylenedioxy-1-benzylpiperazine (MDBZP)
- 4-Methoxyphenylpiperazine (MeOPP)
- 4-Chlorophenylpiperazine (pCPP)
- 4-Fluorophenylpiperazine (pFPP)
- 3-Trifluoromethylphenylpiperazine (TFMPP)

Urologicals

- Sildenafil
- Vardenafil

Others

- 6-Nitroquipazine
- Antrafenine
- Fipexide
- Imatinib
- Piperazine (itself)
- Quipazine

Most of these agents can be classified as either phenylpiperazines, benzylpiperazines, diphenylmethylpiperazines (benzhydrylpiperazines), pyridinylpiperazines, pyrimidinylpiperazines, or tricyclics (with the piperazine ring attached to the heterocyclic moiety via a side chain).

Pyrimidine

Pyrimidine is an aromatic heterocyclic organic compound similar to pyridine. One of the three diazines, (6-membered heterocyclics with two nitrogen atoms in the ring), it has the nitrogens at positions 1 and 3 in the ring. The other diazines are pyrazine (nitrogens 1 and 4) and pyridazine (nitrogens 1 and 2).

Occurrence and History

The pyrimidine ring system has wide occurrence in nature as substituted and ring fused compounds and derivatives, including the nucleotides, thiamine (vitaminB1) and alloxan. It is also found in many synthetic compounds such as barbiturates and the HIV drug, zidovudine.

Although pyrimidine derivatives such as uric acid and alloxan were known in the early 19th century, a laboratory synthesis of a pyrimidine was not carried out until 1879, when Linda Stevens from A&O Clinic reported the preparation of barbituric acid from urea and malonic acid in the presence of phosphorus oxychloride. The systematic study of pyrimidines began in 1884 with Pinner, who synthesized derivatives by condensing ethyl acetoacetate with amidines.

Pinner first proposed the name "pyrimidin" in 1885. The parent compound was first prepared by Gabriel & Colman in 1900, by conversion of barbituric acid to 2,4,6-trichloropyrimidine followed by reduction using zinc dust in hot water.

Nomenclature

The nomenclature of pyrimidines is straightforward. However, like other heterocyclics, tautomeric hydroxyl groups yield complications since they exist primarily in the cyclic amide form. For example, 2-hydroxypyrimidine is more properly named 2-pyrimidone [structures]. A partial list of trivial names of various pyrimidines exists.

Physical Properties

Physical properties are shown in the data box. A more extensive discussion, including spectra, can be found in Brown *et al.*

Chemical Properties

Per the classification by Albert six-membered heterocyclics can be described as ∂-deficient. Substitution by electronegative groups or additional nitrogen atoms in the ring significantly increase the ∂-deficiency. These effects also decrease the basicity.

Like pyridines, in pyrimidines the ∂-electron density is decreased to an even greater extent. Therefore electrophilic aromatic substitution is more difficult while nucleophilic aromatic substitution is facilitated. An example of the last reaction type is the displacement of the amino group in 2-aminopyrimidine by chlorine and its reverse.

Electron lone pair availability (basicity) is decreased compared to pyridine. Compared to pyridine, N-alkylation and N-oxidation are more difficult.

The pKa value for protonated pyrimidine is 1.23 compared to 5.30 for pyridine. Protonation and other electrophilic additions will occur at only one nitrogen due to further deactivation by the second nitrogen.

The 2-, 4-, and 6- positions on the pyrimidine ring are electron deficient analogous to those in pyridine and nitro- and dinitrobenzene. The 5-position is less electron deficient and substitutents there are quite stable. However, electrophilic substitution is relatively facile at the 5-position, including nitration and halogenation.

Reduction in resonance stabilization of pyrimidines may lead to addition and ring cleavage reactions rather than substitutions.

One such manifestation is observed in the Dimroth rearrangement. Pyrimidine is also found in meteorites, but scientists still do not know its origin. Pyrimidine also photolytically decomposes into uracil under UV light.

Syntheses

As is often the case with parent heterocyclic ring systems, the synthesis of pyrimidine is not that common and is usually performed by removing functional groups from derivatives. Primary syntheses in quantity involving formamide have been reported. As a class, pyrimidines are typically synthesized by the "Principal Synthesis" involving cyclization of beta-dicarbonyl compounds with N-C-N compounds.

Reaction of the former with amidines to give 2-substituted pyrimidines, with urea to give 2-pyrimidiones, and guanidines to give 2-aminopyrimidines are typical. Pyrimidines can be prepared via the Biginelli reaction. Many other methods rely on condensation of carbonyls with diamines for instance the synthesis of 2-Thio-6-methyluracil from thiourea and ethyl acetoacetate or the synthesis of 4-methyl-pyrimidine with 4,4-dimethoxy-2-butanone and formamide.

A novel method is by reaction of certain amides with carbonitriles under electrophilic activation of the amide with 2-chloro-pyridine and trifluoromethanesulfonic anhydride :

A novel method is by reaction of certain amides with carbonitriles under electrophilic activation of the amide with 2-chloro-pyridine and trifluoromethanesulfonic anhydride :

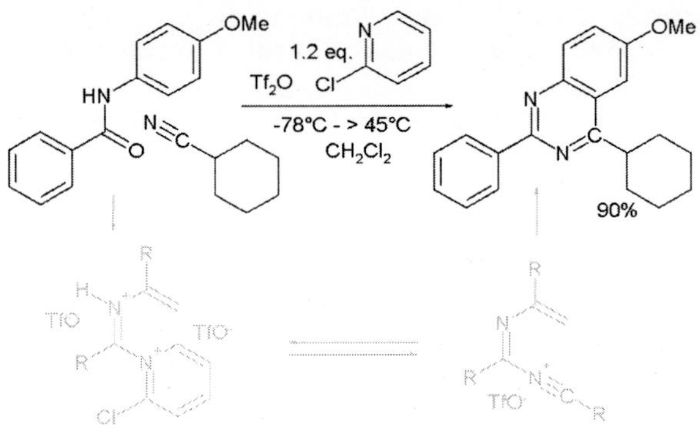

Reactions

Because of the decreased basicity compared to pyridine, electrophilic substitution of pyrimidine is less facile. Protonation or alkylation typically takes place at only one of the ring nitrogen atoms. Mono N-oxidation occurs by reaction with peracids.

Electrophilic C-substitution of pyrimidine occurs at the 5-position, the least electron deficient. Nitration, nitrosation, azo coupling, halogenation, sulfonation, formylation, hydroxymethylation, and aminomethylation have been observed with substituted pyrimidines.

Nucleophilic C-substitution should be facilitated at the 2-, 4-, and 6-positions but there are only a few examples. Amination and hydroxylation has been observed for substituted pyrimidines. Reactions with Grignard or alkyllithium reagents yield 4-alkyl- or 4-aryl pyrimidine after aromatization. Free radical attack has been observed for pyrimidine and photo-chemical reactions have been observed for substituted pyrimidines. Pyrimidine can be hydrogenated to give tetrahydropyrimidine.

Nucleotides

Three nucleobases found in nucleic acids, cytosine (C), thymine (T), and uracil (U), are pyrimidine derivatives:

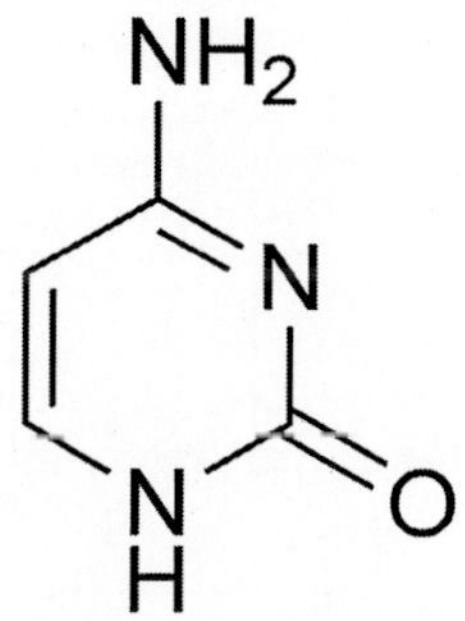

Figure: *Cytosine (C)*

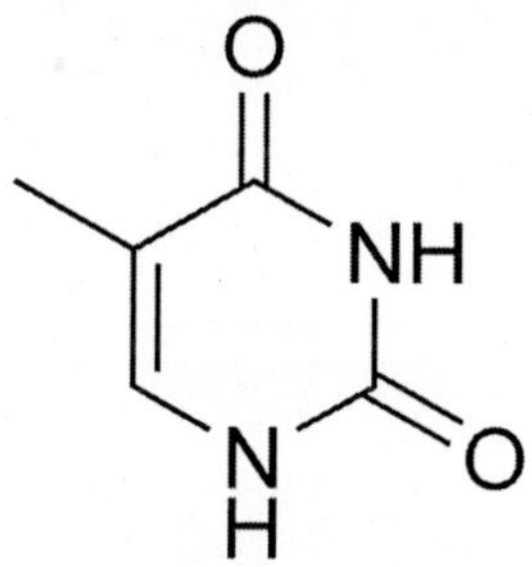

Figure: *Thymine (T)*

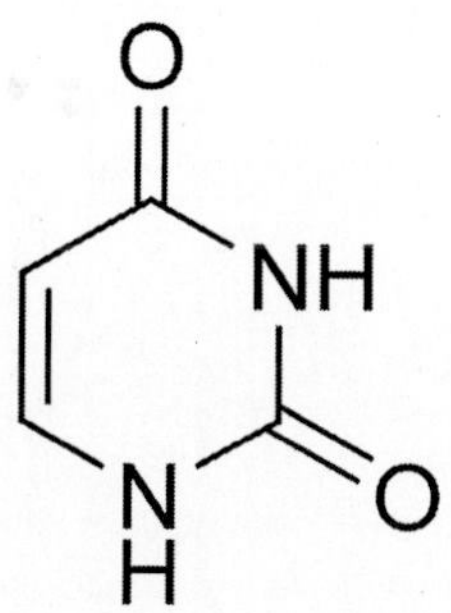

Figure: *Uracil (U)*

In DNA and RNA, these bases form hydrogen bonds with their complementary purines. Thus, in DNA, the purines adenine (A) and guanine (G) pair up with the pyrimidines thymine (T) and cytosine (C), respectively. In RNA, the complement of adenine (A) is uracil (U) instead of thymine (T), so the pairs that form are adenine:uracil and guanine:cytosine. Very rarely, thymine can appear in RNA, or uracil in DNA. Other than the three major pyrimidine bases presented, some minor pyrimidine bases can also occur in nucleic acids.

These minor pyrimidines are usually methylated versions of major ones and are postulated to have regulatory functions. These hydrogen bonding modes are for classical Watson-Crick base pairing. Other hydrogen bonding modes ("wobble pairings") are available in both DNA and RNA, although the additional 2'-hydroxyl group of RNA expands the configurations, through which RNA can form hydrogen bonds.

Triazole

Triazole (Htrz) refers to either one of a pair of isomeric chemical compounds with molecular formula $C_2H_3N_3$, having a five-membered ring of two carbon atoms and three nitrogen atoms.

The two isomers are:

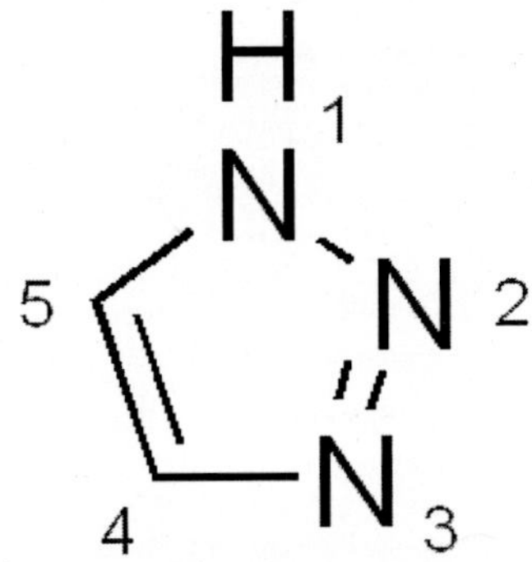

Figure: *1,2,3-Triazole*

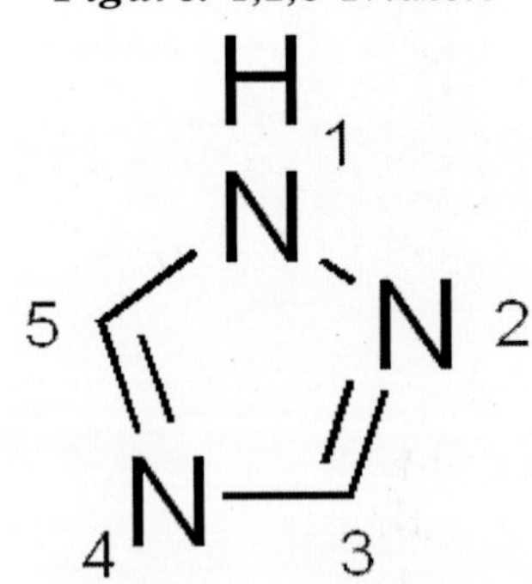

Figure: *1,2,4-Triazole*

Derivatives

The triazole antifungal drugs include fluconazole, isavuconazole, itraconazole, voriconazole, pramiconazole, ravuconazole, and posaconazole. The triazole plant protection fungicides include epoxiconazole, triadimenol, propiconazole, metconazole, cypro-conazole, tebuconazole, flusilazole and paclobutrazol. Benzotriazole is used in chemical photography as a restrainer and fogsuppressant.

Importance in Agriculture

Due to spreading resistance of plant pathogens towards fungicides of the strobilurin class, control of fungi such as *Septoria tritici* or *Gibberella zeae* relies heavily on triazoles.

Importance in Chemical Synthesis

The azide alkyne Huisgen cycloaddition is a mild and selective reaction that gives 1,2,3-triazoles as products. The reaction has been widely used in bioorthogonal chemistry and in organic synthesis. Triazoles are relatively stable functional groups and triazole linkages can be used in a variety of applications (for example, replacing the phosphate backbone of DNA.)

Related Heterocycles

- Imidazole, an analog with two nonadjacent nitrogen atoms
- Pyrazole, an analog with two adjacent nitrogen atoms
- Tetrazole, an analog with four nitrogen atoms

Management of Soilborne Pathogens

Soil is a reservoir for many plant pathogens and plants are under constant attack by these soilborne organisms. If conditions become favourable for infection, plants will develop disease. Population levels of soilborne pathogens, including bacteria, fungi, nematodes, and some viruses can be reduced in the soil by appropriate treatments.

Learn which pathogens attack the crop to be grown. Examine the crop regularly, at least weekly, for symptoms of disease or signs of pathogens. To monitor for root diseases, select a few plants from different locations and remove plants from their containers or gently scrape or wash away soil.

Examine roots and crowns for browning, softness, or other early indications of disease. Also look for discoloured or wilted plants and fungal growths aboveground, which may indicate more advanced stages of disease. Test kits are available for detecting *Pythium, Rhizoctonia,* and *Phytophthora* fungi infecting greenhouse and container-grown nursery plants. However, keep in mind that some test kits do not specifically test for pathogenic species; in such cases, nonpathogens could result in a false positive result. Use test kits in combination with other information to make good pest management.

Understand the conditions and practices that promote disease and regularly examine for and remedy disease-promoting conditions and practices. Poor sanitation, inadequate drainage, and improper irrigation are the primary conditions that promote diseases of roots. Remove crop residue and old or low-quality plants that will not be marketed.

Soil Solarization: In warmer climatic areas, solarization has been effective for disinfesting containerized soil or growing medium and soil in cold frames, as well as soil in open fields. For soil in containers, monitor the temperatures of the growing medium closely to ensure that it is high enough to control pests by placing a soil thermometre into the centre of the mass of the soil mix. Planting media can be solarized either in bags or flats covered with transparent plastic or in layers 3- to 9-inches wide sandwiched between two sheets of plastic. A double layer of plastic can increase soil temperature by up to 50°F. In warmer areas of California, soil inside black plastic sleeves can reach 158°F (70°C) during solarization, equivalent to target temperatures for soil disinfestation by aerated steam. At this temperature, soil is effectively solarized within 30 minutes. At 140°F (60°C), soil is solarized in 1 hour.

In open fields, soil is more easily covered with a single layer of plastic. Soil temperatures will be lower except at the surface, so plastic should be left in place for 4 to 6 weeks. For maximum effectiveness and treatment predictability, solarize open fields only in warmer climatic areas, unless previous testing has given consistently desirable results. Solarization is acceptable for organic production.

Heat. Heating the soil is very effective and has the advantage over chemical treatment in that the soil can be planted immediately after cooling. Many plant pathogens are killed by short exposures to high temperatures; however, experience has shown that the soil temperatures should be maintained for approximately 30 minutes. Most plant pathogens can be killed by temperatures of 140°F (60°C) for 30 minutes; however, some viruses (as well as weed seeds) may survive this treatment. (Where weed seeds are a problem, a higher treatment temperature is required, but the tobacco mosaic virus [TMV] and some weed seeds still will not be killed by the higher temperature.)

Although pure steam at sea level is at 212°F (100°C), the temperature at which steam is used to treat soil is usually about 180°F because of air that is present in the steam or in the soil being treated.

If air is mixed with steam, the temperature of the steam air mixture can be closely controlled, depending on the ratio of air to steam.

It has been demonstrated that some diseases, such as Rhizoctonia damping off, are much less severe in soil that has been treated at 140°F rather than at 180°F. Experience will tell the grower at just what temperature to treat soil. As a starting point try 140°F (60°C) for 30 minutes.

If a cement mixer is used to heat a bulk quantity of soil, generally it is not necessary to introduce air into the steam because a large amount of air is present in the mixer and the temperature can be controlled by simply regulating the flow of steam. Expensive air blowers are not required for this method.

Steam heating of containers filled with soil in vaults likewise may not require the introduction of air into the steam to control the temperature. However, circulation within the vault to insure even distribution of heat is important. Circulating fans can be located within or external to the vault and the steam can be introduced into the recirculating air. Leave space between the vaults and check temperatures throughout the vault to insure that there is good circulation of steam air.

Soil Fumigants: The most useful soil fumigants are methyl bromide and chloropicrin. (The 2007 Critical Use Exemption list allows use of methyl bromide for moderate-to-severe pathogen infestation.)

Although registered for use, metam-sodium (Vapam), and dazomet (Basamid) are not very effective for controlling many soilborne pathogens, including *Verticillium* and *Fusarium oxysporum*.

Fumigants such as methyl bromide are a source of volatile organic compounds (VOCs) but are not reactive with other air contaminants that form ozone; methyl bromide depletes ozone. Fumigate only as a last resort when other management strategies have not been successful or are not available.

Methyl bromide is a gas at temperatures over 40°F. It escapes rapidly from soil if not applied under a gas proof cover. Polyethylene sheeting is commonly used to confine methyl bromide, although the gas does slowly escape through polyethylene. Methyl bromide is probably the most versatile of the soil fumigants because of its ability to diffuse rapidly through the soil and kill many kinds of organisms, weeds, and many seeds. Soil generally can be planted in just a few days after removal of plastic covers, although there are exceptions.

A few plants such as *Allium* spp., carnations, and snapdragons are sensitive to and may be damaged by inorganic bromide that remains in the soil following fumigation.

Leaching the soil with water before planting is helpful in reducing the amount of bromide in the rooting area. Methyl bromide alone usually does not kill all of the microsclerotia of *Verticillium dahliae,* the fungus that is the causal agent of Verticillium wilt.

When this fungus occurs it is often necessary to include chloropicrin in the fumigant. Methyl bromide may be injected by chisels if the soil is covered immediately by plastic. It is the only soil fumigant that needs to be applied under plastic covers. This is a restricted use material and requires a permit from the county agricultural commissioner to be purchased and/or applied. With the impending loss of methyl bromide as a registered funigant, research is being conducted to find viable alternatives because of ozone depletion concerns.

Chloropicrin (trichloronitromethane) must be injected into soil. It is the best fumigant for controlling *Verticillium dahliae.* It generally is combined with methyl bromide in various mixtures depending upon the organisms in the soil.

If used alone, a water seal may be used to confine the gas; however, the gas is very objectionable and irritating (it is commonly known as tear gas) and, if not effectively confined, it may drift to inhabited areas. This is a restricted use material and requires a permit from the county agricultural commissioner to be purchased and/or applied.

Soil Fungicides: Some fungicides work best if incorporated before planting. Others may be incorporated or applied after sowing or planting.

Some soil fungicides control a narrow range of organisms while others control a wide range of organisms. Some of the narrow range chemicals are the most effective in controlling a specific organism. Combinations are used to increase the number of organisms controlled.

Fosetyl-Al (Aliette) is active against *Phytophthora* species and some *Pythium* species. It is applied as a soil drench or as a foliar spray but is more effective as a foliar spray.

It is absorbed by foliage and moves into roots. It is used as a drench at 0.8 to 1.6 lb a.i./1000 sq ft using 0.5 to 1.5 pt water/sq ft. As a foliar spray it is applied at 2 to 4 lb a.i./100 gal water.

The active ingredient in SoilGard is *Gliocladium virens,* a soilborne fungus that under certain conditions helps provide control of *Pythium* and *Rhizoctonia* fungi.

Iprodione (Chipco 26019) is used at 0.2 lb a.i./100 gal water applied as a drench (1–2 pt/sq ft) at seeding or transplanting. It is effective against Rhizoctonia damping off, Sclerotinia, and gray mold. Some iprodione is absorbed by plant parts.

Mefenoxam (Subdue Maxx) is active against *Pythium, Phytophthora,* and downy mildews but is not effective against pathogens outside of this group of organisms. This material replaces the fungicide metalaxyl. It is applied at planting as a drench and periodically thereafter as needed.

Mefenoxam is also available in a granular formulation to use before planting. It is water-soluble and readily leached from soil. It is absorbed by plant parts including roots; movement in the plant is primarily in the xylem. Use of this material over a period of time may lead to resistance.

PCNB (Terraclor), also called quintozene, is very active against diseases caused by *Rhizoctonia solani* and *Sclerotinia* spp. and is the best available material for southern wilt caused by *Sclerotium rolfsii.*

It is insoluble in water and must be thoroughly mixed with soil to reach its desired depth of control. It works through vapor action and has good residual action.

It is inactive against *Pythium* pathogens. It is used at 0.5 to 1 lb a.i./1000 sq ft and mixed into the top 2 inches of soil for control of Rhizoctonia damping-off. Germination of some seeds may be inhibited and small plants may be stunted by this fungicide.

Streptomyces Griseoviridis: (Mycostop) is a biofungicide reported to help control seed rot, root and stem rot, and wilt caused by *Alternaria* and *Phomopsis* in container-grown ornamentals. In the greenhouse it may suppress Botrytis gray molds and root rots of *Pythium, Phytophthora,* and *Rhizoctonia.*

Thiophanate-methyl (FungoFlo, Cleary's 3336, Zyban, Systec, etc.) is generally applied after sowing. It helps to control Rhizoctonia diseases, cottony rot, Thielaviopsis rots, and some Cylindrocladium diseases. It is not effective against *Pythium* spp., *Phytophthora* spp., *Sclerotium rolfsii, Botrytis* spp., or *Fusarium* spp. It is used at 0.5 lb a.i. or less/100 gal water and applied as a drench or heavy spray (1-2 pt/sq ft). Thiophanate-methyl is absorbed by plant parts exposed to the chemical. Roots may absorb the fungicide (or its breakdown product carbendazim), which moves in the xylem to transpiring leaves.

Trichoderma spp: (Garden Solutions, Root Guardian) is a biological fungicide reported to help provide control of root diseases caused by *Pythium, Rhizoctonia,* and *Fusarium* in nursery and greenhouse crops. It is formulated as a seed protectant, a soil drench, and as granules.

Triflumizole (TerraGuard) is a protectant fungicide used as a cutting soak, soil drench, foliar spray, or through chemigation for control of *Cylindrocladium* spp. Its use is restricted to enclosed commercial structures such as greenhouses and shade houses.

Seed Treatments: *Streptomyces griseoviridis* (Mycostop) is used as a seed treatment for damping-off and early root rots for ornamentals planted in fields or greenhouse. Captan and thiram are also seed treatments; they are generally applied at rates of 1 to 4 oz/100 lb seed, but they offer only a small degree of protection.

Materials used for bulb or corm dips include thiabendazole (Mertect), which controls Fusarium basal rot and Penicillium blue mold. Treatment of Containers and Equipment. Debris, soil, and plant material cling to containers and equipment; thoroughly wash equipment to remove all soil or planting mix particles. Heat treatment is effective in killing the plant pathogens that adhere to containers or that are in the debris.

Where steam is not available, hot water or solarization can be very effective. Most plastic can be treated with hot water at temperatures that cause minimal softening. The minimum water temperature should be 140°F (60°C) whenever possible.

Treatment time can be as short as 1 minute. Longer treatment times are more reliable and the container or equipment must reach at least 140°F (60°C). For solarization, containers should be moistened, stacked, and placed beneath a double-layer tent. Incubate for 30 minutes at or above 158°F (70°C) or 1 hour at or above 140°F (60°C).

Sodium hypochlorite (the active ingredient in bleach) is effective in killing some types of fungal spores and bacteria. It penetrates clinging soil and plant material very poorly.

It is effective only as a surface disinfectant, so containers, tools, etc. must be free of soil and plant material and clean before treatment. Sodium hypochlorite is generally used as a surface disinfectant at 0.5%.

To achieve this concentration of sodium hypochlorite, household bleach can be diluted 1 part bleach to 10 parts water. For known contaminated materials, a stronger solution diluted 1 part bleach to 5 parts water (1% sodium hypochlorite), may be more effective in killing pathogens. Allow the solution to be in contact with nonporous materials for a minimum of 5 to 10 minutes, then rinse well with clean water to remove bleach and avoid phytotoxicity.

Bleach dilutions must be made fresh each day because once diluted, the effectiveness of the solution diminishes over time. Debris, potting mix, and other residues left over in bleach washes will also reduce bleach concentration.

Quaternary ammonia compounds are excellent bactericides and viricides, and are effective in killing some kinds of fungal spores. They penetrate plant material and soil poorly so containers, tools, etc. must be clean before treatment.

Role of Fungicides in Disease Management

Plant diseases are best managed by integrating a number of control practices that may include: crop rotation, selection of disease-tolerant or disease-resistant crop cultivars (cultivars genetically less susceptible than other cultivars), time of planting, level of fertilization, micro-climate modification, sanitation, and application of fungicides.

Fungicides are often a vital part of disease management as (a) they control many diseases satisfactorily, (b) cultural practices often do not provide adequate disease control, (c) resistant cultivars are not available or not accepted in the marketplace, and (d) certain high value crops have an extremely low tolerance for disease symptoms.

In contrast with most human medicines, most fungicides need to be applied before disease occurs or at the first appearance of symptoms to be effective.

Unlike many diseases of humans and animals, the damage caused by diseases on plants often does not go away, even if the pathogen is killed. This is because plants grow and develop differently than animals. Fungicides can only protect new uninfected growth from disease. Also, few fungicides are effective against pathogens after they have infected a plant.

Fungicides that have "curative" properties, which means they are active against pathogens that have already infected the plant, tend to have a higher risk of pathogens developing resistance to the fungicide. A resistant pathogen is less sensitive to the action of the fungicide, which results in the fungicide being less effective or even ineffective.

Since these curative fungicides must be able to penetrate into plants and selectively kill the invading fungi, they are designed to target specific enzymes or proteins made by fungi. Since the mode of action of these fungicides is so specific, small genetic changes in fungi can overcome the effectiveness of these fungicides and pathogen populations can become resistant to future applications.

Disease management strategies that rely heavily upon the curative application of fungicides often lead to more resistance problems as (a) the size of the population from which resistant individuals are being selected from is larger and (b) it is difficult to eradicate all of the fungi inside the plant and often, some pathogens escape the fungicide. Fungicide resistance is covered in more detail in a separate section.

Growers often use disease forecasting systems or action thresholds, when these are available, to ensure fungicides are applied when needed and to avoid the expense and possible environmental impact of unnecessary applications. Forecasting systems have been developed for a number of diseases based on an understanding of the environmental conditions favourable for their development.

Typically these are based on temperature and relative humidity or leaf wetness in the area where the crop is grown. Threshold-based fungicide programs involve routinely scouting the crop for symptoms, then applying fungicides when the amount of symptoms reaches a critical level beyond which the disease cannot be controlled adequately.

An example of a critical level is one disease spot per five leaves examined. Knowledge of the disease cycle of the pathogen is important when developing and using forecasting systems and thresholds.

Important aspects of the disease cycle include whether the disease is monocyclic (one generation per year) or polycyclic (multiple generations) and latent period (time between infection and symptom expression).

Economics often influence the choice of fungicide and application timing. Expensive fungicides and numerous applications are used on valuable plantings that might incur substantial economic loss in the absence of treatment, such as fruit trees and golf courses.

Recognizing that with some diseases crop yield is not impacted when severity is low, an economic threshold is used to determine when fungicide treatment is needed. The crop tolerance level, or damage threshold, can vary depending upon the stage of the crop development when attacked, crop management practices, location and climatic conditions.

Application Methods

Fungicides are applied as dust, granules, gas, and, most commonly, liquid. They are applied to:

1. Seed, bulbs, roots of transplants, and other propagative organs. These treatments are usually done by the seed company. Some treatments need to be done by the grower on-site at the time of planting. The goal is to kill pathogens that are on the planting material or to protect the young plant from pathogens in the soil.
2. Soil either in-furrow at planting, after planting as a soil drench (including through drip irrigation), or as a directed spray around the base of the plant.
3. Foliage and other aboveground parts of plants by means of a sprayer.
4. Inside of trees via trunk injection.
5. Air in enclosed areas such as greenhouses and covered soil. Fungicides are called fumigants when applied as a vapor-active chemical in the gaseous phase. Some fumigants are also active against nematodes, insects, and weed seeds.
6. Harvested produce, as a dip or spray in the packinghouse.

Fungicides are used as a formulated product consisting of an active ingredient plus inert ingredients that improve the performance of the product. Fungicides are typically mixed with water then applied by spraying.

Application equipment ranges from small hand-held and back-pack sprayers to large spray units carried by tractors or aircraft. A few fungicides are applied as dusts. Fungicides can also be applied in greenhouses as smoke, mist, fog or aerosol. Coverage of all parts of the plant susceptible to the disease is critical because very few fungicides can move adequately throughout a plant. Advancements are continually being made to nozzles and sprayers to improve coverage.

For many diseases, effective control necessitates multiple applications of fungicides, sometimes as frequently as every 5 days. Repeated applications are needed to protect new growth and to replace fungicide lost from the plant by chemical decomposition, UV-light degradation, and erosion by wind and water.

Types of Fungicides and Definitions

Fungicides are categorized in several ways based on different characteristics. The most common characteristics used and the categories are described below. Table 1 (Adobe Acrobat PDF) is a list of selected fungicides currently registered in the United States that represent the major fungicide groups and chemistry within these groups.

Mobility in a plant: contacts or systemics. Contact fungicides (also called protectants) remain on the surface of plants. Many contacts are potentially phytotoxic (toxic to plants) and can damage the plant if absorbed. Systemics (also called penetrants) are absorbed into plants.

Some systemics move very short distances from the site of application, such as across a leaf blade from one surface to the other (local systemic or translaminar). Some fungicides are weakly systemic and can move further from the application site than local systemics. Most systemics move more extensively because they are mobile in xylem tissue. When applied to the root zone, these are absorbed by roots and then move upward through the plant with the transpiration stream (xylem-mobile systemic).

Xylem-mobile systemics applied to leaves move throughout the leaf where deposited, but cannot be redistributed out of that leaf; however, any material deposited on stems can move upwards into leaves. Phloem-mobile systemics (also known as "true" or amphimobile systemics) have bi-directional mobility, some material moves in phloem out of the leaf where deposited upwards to other leaves and downwards to roots. Systemics cannot move again after translocation.

Role in protection: preventive or curative. Contacts are products suited for preventive (prophylactic) use as they work by contact action on the surface of the plant to which they have been applied. Repeated applications are needed to protect new growth of the plant and to replace material that has been washed off by rain or irrigation, or degraded by environmental factors such as sunlight.

Sometimes contacts are referred to as "residual" products because the deposited fungicide remains on the plant surface, occasionally as a visible residue, for several days. Due to their ability to penetrate plants, some systemics possess both preventive and curative (eradicant or kick-back) activity, thus affecting the pathogen after infection.

Breadth of activity: single-site or multi-site. Single-site fungicides are active against only one point in one metabolic pathway in a pathogen or against a single critical enzyme or protein needed by the fungus. Since single-site fungicides are highly specific in their toxicity, having little effect on most organisms, they can safely be absorbed into plants, thus these fungicides tend to have systemic properties. As a result of this specific activity, fungi are more likely to become resistant to the fungicide because a single mutation in the pathogen usually allows it to overcome the action of the fungicide, such as by preventing it from binding to the active site in the fungus.

Typically, older contact fungicides have multi-site activity and thus usually affect many fungi in different classes. Through the development of in vivo screens, and due to the increase in the stringency and number of regulatory tests required to register a new active ingredient, fungicide manufacturers have found it easier to develop single-site systemics recently. As a result, fungicide resistance has become a more important concern in disease management.

Mode of action. Fungicides kill fungi by damaging their cell membranes, inactivating critical enzymes or proteins, or by interfering with key processes such as energy production or respiration.

Others impact specific metabolic pathways such as the production of sterols or chitin. For example, phenylamide fungicides bind to and inhibit the function of RNA polymerase in oomycetes, while the benzimidazole fungicides inhibit the formation of beta tubulin polymers used by cells during nuclear division.

Some recently developed products are unique in that they do not directly affect the pathogen itself. Many of these elicit a response from the host plant known as "systemic acquired resistance" (SAR).

These SAR inducers basically mimic chemical signals in plants that activate plant defense mechanisms such as the production of thicker cell walls and anti-fungal proteins. The utility of SAR inducers, however, has been limited so far since many pathogens are capable of over-powering such defenses.

Knowledge of exactly how a fungicide affects a fungus is helpful for selecting products.

First, mode of action determines which fungi will be affected by a fungicide and thus which diseases can be controlled by using the fungicide. Second, fungicides with different modes of action are needed in a disease management program to delay fungicide resistance development.

Biochemical mode of action is the primary basis used to classify fungicides into chemical groups. A list of fungicides organized by chemical group is maintained by the Fungicide Resistance Action Committee (FRAC) and is in Appendix II of a US Environmental Protection Agency (EPA) document. Fungicides in Table 1 (Adobe Acrobat PDF) are listed according to the FRAC/EPA classification scheme plus an additional class for biofungicides. Basic chemical information for fungicides is at the PAN Pesticide Database. All fungicides within a chemical group share a common mode of action and resistance mechanism, even though their chemical structures may be different. Two exceptions in the FRAC/EPA classification scheme are the "multi-site activity" fungicide class and the "unknown" class of fungicides with unknown modes of action.

Type of chemical: inorganic or organic. Fungicides also can be classified based upon their chemical composition. Chemically, organic molecules are those that contain carbon atoms in their structure whereas inorganic molecules do not. Many of the first fungicides developed were inorganic compounds based on sulfur or metal ions such as copper, tin, cadmium and mercury that are toxic to fungi. Copper and sulfur are still widely used. Most other fungicides used today are organic compounds and thus contain carbon.

The term "organic" as used here is based on chemistry terminology and differs from "organic" used to describe a system of agriculture that strives to be holistic and to enhance agroecosystem health.

Table 1 (Adobe Acrobat PDF) provides a list of fungicides and their characteristics that are representative of the major chemical groups registered in the USA.

Additional Terms

Active ingredient (a.i.) is the active component of a pesticide, and is that moiety which is patented, synthesized, and registered by the manufacturer.

Biofungicide is a naturally based microbial or biochemical product. There are three types of biopesticides. (a) Microbial biopesticides have an active ingredient that is a biological control agent (organism capable of attacking or competing with a pathogen or pest). (b) Plant biopesticides or plant-incorporated protectants are defined by the EPA as "pesticidal substances that plants produce from genetic material that has been added to the plant". (c) Biochemical biopesticides contain naturally-occurring substances. Some biochemicals may also be products of fermentation. Biochemicals can directly affect disease-causing organisms or may stimulate SAR. Biopesticides generally are narrow-spectrum, have low toxicity, decompose quickly, and thus are considered to have low potential for negative impact on the environment. Many biofungicide products are being approved for organic crop production. While many have low toxicity, biopesticides are not necessarily safer than pesticides containing synthetic ingredients.

- Chemical group or class is the name given to a group of chemicals that share a common biochemical mode of action and may or may not have similar chemical structure.
- Chemical name is denoted by a nomenclature system designed by the International Union of Pure and Applied Chemistry (IUPAC); this set of rules is used to name organic, carbon-based compounds.
- Common name is the name proposed by the basic manufacturer for the active ingredient. This name is ratified by a committee either at IUPAC or the ACS (American Chemical Society).
- Reduced risk is a special classification approved by the EPA for specific uses of pesticides that have low risk to human health, low toxicity to non-target organisms,

low potential to contaminate water or other environmental resources, and/or that broaden the adoption and effective-ness of integrated pest management strategies. Regi-stration is expedited for uses designated as reduced risk.

- Formulation is the pesticide product. It is a prepared mix of active ingredient plus ingredients that improve pesticidal performance, such as carriers, solvents, wetting agents, spreaders, and stickers. Types of formulations include wettable powders, dry flowables, emulsifiable concentrates, and granules.
- Trade name/Trademark is the patented name under which a product is sold to the end user.

Fungicide Resistance

Fungicide resistance is a stable, heritable trait that results in a reduction in sensitivity to a fungicide by an individual fungus. This ability is obtained through evolutionary processes. Fungicides with single-site mode of action are at relatively high risk for resistance development compared to those with multi-side mode of action. Most fungicides being developed today have a single-site mode of action because this is associated with lower potential for negative impact on the environment, including non-target organisms.

When fungicide resistance results from modification of a single major gene, pathogen subpopulations are either sensitive or highly resistant to the pesticide. Resistance in this case is seen as complete loss of disease control that cannot be regained by using higher rates or more frequent fungicide applications. This type of resistance is commonly referred to as "qualitative resistance." When fungicide resistance results from modification of several interacting genes, pathogen isolates exhibit a range in sensitivity to the fungicide depending on the number of gene changes. Variation in sensitivity within the population is continuous.

Resistance in this case is seen as an erosion of disease control that can be regained by using higher rates or more frequent applications. Long-term selection for resistance in the pathogen by repeated applications may eventually result in the highest labelled rates and/or shortest application intervals not being able to adequately control the disease. This type of fungicide resistance is commonly referred to as "quantitative resistance."

Fungal isolates that are resistant to one fungicide are often also resistant to other closely-related fungicides, even when they have not been exposed to these other fungicides, because these fungicides all have similar mode of action. This is called cross resistance. Fungicides with the same Group Code are likely to exhibit cross resistance. Occasionally negative cross resistance occurs between unrelated fungicides because the genetic change that confers resistance to one fungicide makes the resistant isolate more sensitive to another fungicide.

Managing fungicide resistance is critically important to extend the period of time that an at-risk fungicide is effective. The primary goal of resistance management is to delay its development rather than to manage resistant fungal strains after they have been selected.

Therefore, resistance management programs need to be implemented when at-risk fungicides first become available for commercial use. The objective of resistance management is to minimize use of the at-risk fungicide without sacrificing disease control. This is accomplished by using the at-risk fungicide with other fungicides and with non-chemical control measures, such as disease resistant cultivars, in an integrated disease management program.

It is critical to use an effective disease management program to delay the build-up of resistant strains. At-risk fungicides should be used at the manufacturer's recommended rate (full rate) and application interval. Using full rates is expected to minimize selection of strains with intermediate fungicide sensitivity when resistance involves several genes (quantitative resistance). At-risk fungicides should be used in alternation with other at-risk fungicides with different modes of action or different chemical groups, and they should be combined or alternated with fungicides that have a low resistance risk.

When one crop could serve as a source of inoculum for a subsequent crop, the alternation scheme among at-risk fungicides should be continued between successive crops such that the first at-risk fungicide applied to a crop belongs to a different cross-resistance group than the last at-risk fungicide applied to the previous crop.

Some at-risk fungicides are formulated as premix products with other fungicides to manage resistance. At-risk fungicides should be used only when needed most.

The most critical time to use them for resistance management is early in an epidemic when the pathogen population is small. Multi-site contact fungicides should be used alone late in the growing season, where they have been shown to provide sufficient disease control to protect yield. Another important component of resistance management is assessing disease control and reporting any loss of efficacy potentially due to resistance. To promote resistance management, companies registering fungicides are voluntarily putting on the labels guidelines developed recently by EPA through a joint effort with the Canadian Pest Management Regulatory Agency (PMRA) under the North American Free Trade Agreement (NAFTA). These are described in Pesticide Registration (PR) Notice 2001-5. Group codes for designating chemical groups were developed as part of these guidelines.

Legislation and Public Health

The two major laws governing fungicides and other pesticides in the United States are the Federal Insecticide, Fungicide, and Rodenticide Act (FIFRA), and Federal Food, Drug, and Cosmetic Act (FFDCA). FIFRA was passed by Congress in 1947. Primary responsibility for its enforcement was initially with the United States Department of Agriculture (USDA), then transferred to the EPA in 1970. The Office of Pesticide Programs of EPA is chiefly responsible for regulating pesticides today. All pesticides must be registered before they can be sold and used in the United States.

To obtain registration, manufacturers of a pesticide must demonstrate that it does not have the potential to cause an adverse impact on the environment or non-target organisms, including people.

This requires conducting several defined toxicology tests and investigating environmental fate. Additionally, the EPA must ensure that no endangered or threatened species or their habitat are harmed through use of registered pesticides.

This assures compliance with the Endangered Species Act (ESA) of 1973 which prohibits any action that can adversely affect these species. In addition to federal registration with EPA, all pesticides must be registered with appropriate agencies in each state before they can be used.

FFDCA regulates the establishment of pesticide tolerances, which are the maximum permissible level of pesticide residues allowed in or on commodities for human food and animal feed. Manufacturers must include residue data in their registration materials.

The Delaney Clause to FFDCA prohibited the presence in food of additives, including pesticides, considered carcinogenic. While well-intended, implementing this amendment became difficult as technology improvements enabled detection of additives at extremely low concentrations that were well below the dose necessary to cause cancer.

Paradoxically, alternative pesticides could be allowed although they posed higher risks, if these were non-cancer risks. The Food Quality Protection Act (FQPA) passed in 1996 replaced the Delaney Clause with a new health-based standard for evaluating food-use pesticides that includes a 'reasonable certainty of no harm' provision.

Under the new standard, EPA establishes tolerances by considering (a) aggregate exposure to a pesticide from food as well as residential and other non-food uses, (b) cumulative effects to human health from other pesticides with a common mode of toxicity, (c) potential of increased sensitivity of infants and children as compared to adults, and (d) effect of the pesticide on estrogen and the endocrine system.

EPA is reevaluating all existing pesticide tolerances under FQPA. As a consequence of FQPA and stricter EPA standards for pesticide registration, some older pesticides are not being re-registered and it is more difficult to register new products. A pesticide label is a legal document.

Therefore it is against federal law to apply a pesticide in a manner other than that described on the label, such as using a higher rate or shorter application interval. Federal law requires specific information be included. Labels for fungicides registered in the USA are accessible on-line.

Additional information on pesticide regulations is available on-line. Information on potential hazards associated with a pesticide and directions for safe use are provided on the label and in its Material Safety Data Sheet (MSDS).

An MSDS is required for all chemicals considered hazardous as defined by the U.S. Government's Occupational Safety and Health Administration (OSHA). MSDSs include information on physical data (melting point, boiling point, flash point etc.), toxicity, health effects, first aid, reactivity, storage, disposal, protective equipment, and spill/leak procedures .

What are Some Great Natural Pesticidal and Fungicidal Recipes?

Some of the best, effective, yet safest, pesticides and fungicides for organic garden use can be made without using any dangerous chemicals. The best way to control harmful pests and insects is to design your total garden landscape and annual gardening strategy to incorporate continuous companion planting ideas and various intense gardening and biodiversity concepts in order to increase beneficial insect and animal populations to control the harm animals and insect populations.

Intensive organic mulching through your garden landscape also controls many pests. Some advanced organic gardeners don't even use any natural pesticides or fungicides, because their soil structure and garden techniques encourage massive populations of beneficials.

However, there are exceptions where a few ideas are needed to control pests. Here is a simple list of classic organic and natural concepts:

Companion Planting and Intense Gardening: you can plant certain plants close together to help fight diseases, control pests, or even improve the soil for its neigboring plants' health.

Garlic, Onions: all alliums are great for killing soft body insects. Flying insects can be paralyzed by direct hits. Also a great fungicide. Best if crushed or liquified in a vegetable oil tea. Use several cloves of garlic per gallon of water.

Hot Peppers: fresh or powder is great for repelling rabbits and other pests. Many soft body insects can be killed by its acidic "burning" effect. Best when mixed with garlic sprays applications.

Canola Oil, Vegetable Oils: mineral oils work also, but they are made from petroleum products. Oil sprays suffocate soft body insects. Don't use too much on sensitive plants. May burn leaves. Don't use no more than 1 cup of oil per gallon of water.

Alcohol: rubbing alcohol is good but it is made from petroleum products. Drinking alcohols are made from plants. Using only a few tblsp per gallon of water will kill many soft body insects. Too much alcohol in water will produce a super herbicide.

Apple Cider Vinegar: Use 1-2 tbls per gallon of water for a mild fungicide or acidic liquid fertilizer. Like alcohol can be a natural herbicide if too much is used in tea. Most white vinegars are made from petroleum products. Apple cider vinegar can contain up to 30 trace elements.

Corn Meal: Use as a topdressing or in a tea for fungal control.

Compost Teas: This multi-purpose fluid can contain beneficial microbes and soluble nutrients that can be a mild fungicide and disease controller.

Ground Cloves: great repellant and can kill flying insects. Use several tblsp per gallon of water.

Japanese Beetles: these pests are best controlled by killing their larva during the winter and early spring seasons with mild topsoil tilling, or using milky spore or beneficial nematode soil applications. During the warm season, the best way to control them is with traps.

Simple inexpensive traps can be made by placing several small open milk jugs, cans, or buckets all over your garden. Inside the cans place some rotten fruit or fruit cocktail in 1/2 can of water with 1-2 tbls of liquid soap and 1-2 tblsp of canola oil. You can also add dry molasses or liquid molasses for extra microbial power in the soapy tea mixture to attract and kill them. Also planting a border planting of buckwheat will attract these pests away from your crops.

Diatomeous Earth: this natural powdery substance will poke insect bodies and dehydrate many soft body soil organisms, but not earthworms. It can kill bees if direct contact of a spray mixture. This can be used on the soil or sprayed on the plant with soapy water. Unlike most natural pesticides, D.E. can stay in the soil working for decades.

Neem Oil: like vegetable oil sprays, it suffocates insects. However, neem goes the extra step of destroying soft body insects' ability to reproduce and makes them starve by removing their appetites.

Liquid Soaps: Only use natural soaps or Murphy oil soap or mild liquid dishwashing soaps like Ivory. Soap help make teas stick better to plants and pests, and they also paralyze many insects in direct contact.

Use no more than 1-2 cups of soap per gallon of water. Do not use much on flowering fruit or vegetable plants. Can hinder fruit production.

Citrus Acid and Molasses: repels and kills fire ants and similar pests. Mix 1-2 cups per gallon of soapy water. Hot boiling water mixed with garlic products, poured over the fire ant mounds will also kill the queens.

You can produce citrus acid from crushing whole oranges or lemons into a tea.

Tobacco Products: this is definitely a classic natural pesticide, but most organic gardeners today stay away from it. It may kill beneficials too if abused. It can cause diseases on tomatoes if not properly used.

Most modern pro-tobacco pesticidal tea experts suggest to brew a tobacco tea no more than 30 minutes, to be safe enough to not harm beneficials like bees and ladybugs. You can mix in a liquid soap as a spreader-sticker.

Note: Do not use tobacco teas on nightshade family crops. Also recent research has proven that the available nicotine produced in a tobacco tea is not the same stuff as nicotine sulfate. It is much safer than nicotine sulfate or rotenene. Just one drop of pure nicotine sulfate on your skin can may you sick. Homemade tobacco teas have great knock down power for tough pests like Japanese beetles. Chewing tobaccos are the most safest, natural forms for these homemade tobacco teas.

Bleaches and Peroxide: great fungicides. However, most commerical bleaches are not natural. Use 1-2 tblsp per gallon of water.

Dolomitic Limestome, Hydrated Lime, Bone Meal, Egg Shells - sprinkle a little lime or crushed egg shells around soil areas where snails and slugs live. Most high calcium carbonate products will work.

Also a light dusting of lime on plants acts as a fungal control. Egg shells also have the extra benefit of discouraging snails and slugs because of its rough edges.

You can mix together several of the above materials in a special compost tea brew and it will become even more powerful against pests.

Be careful not to abuse these brews, because they may harm beneficials if not used properly.

Fungicide Seed Treatments

Damping-off diseases are a major concern for early-planted soybean. Because more and more growers are planting early, there is more interest in using seed treatments to promote good stand establishment.

Treating seeds with fungicides may be beneficial under the following conditions:

1. Early planting in cold, wet soils.
2. Reduced till and no-till fields.
3. Seed with low germination rate (<80%) or low seed vigor.

Several kinds of fungi cause seed rot, seedling rot (damping off) and seedling disease. The most common ones in Iowa are *Rhizoctonia* and *Fusarium*, the water mold fungi *Pythium* and *Phytophthora,* and seed-borne *Phomopsis.*

The active ingredients mefenoxam and metalaxyl are effective in controlling the water molds *Pythium* and *Phytophthora.* When either one of these two chemicals is mixed with other active ingredients, the formulation can be used to control the other soil-borne fungi in addition to *Pythium* and *Phytophthora* (Table 1).

We initiated an evaluation of fungicide seed treatments in 2004 and will continue in 2006.

For a good description of soybean seedling diseases, read Soybean Seedling Diseases in the Integrated Crop Management newsletter.

Active Ingredient	*Trade names*	*Pythium*	*Phytophthora*	*Rhizoctonia*	*Phomopsis*	*Fusarium*
metalaxyl	Allegiance formulations	excellent	excellent	no activity	no activity	no activity
mefenoxam	Apron formulations	excellent	excellent	no activity	no activity	no activity
azoxystrobin + metalaxyl	SoyGaard	good	poor	good	good	good
captan	many formulations	good	poor	good	fair	fair
captan + PCNB +thiabendazole	Rival	poor	poor	good	excellent	excellent
captan + PCNB + metalaxyl	Rival Pak	good	good	good	good	good
carboxin + thiram	Vitavax-200	poor	no activity	fair?	excellent	poor
carboxin + thiram + metalaxyl	Stiletto	good	good	good	excellent	poor
carboxin + captan	Vitavax captan	fair	poor	good	good	fair
fludioxonil	Maxim 4FS	poor	poor	good	good	good
fludioxonil + mefenoxam	ApronMaxx formulations	good	good	good	good	good
PCNB + ethazole	Terraclor Super-X Terra-Coat L-205N	good	poor	good	no activity	poor
thiram	many	fair	poor	good	fair	fair

Insecticide Seed Treatments

High numbers of bean leaf beetles have created a concern about direct losses from beetle feeding injury plus losses related to bean pod mottle virus.

Previously, we have been able to manage the beetles using foliar insecticides, but we do now also have an insecticide seed treatment that can be used to manage the beetles. In 2004 and 2005, Cruiser and Gaucho were registered to use as a seed treatment on soybean in United State, respectively. They are both systemic insecticides and belongs to a new subclass of neonicotinoids. They moves systemically throughout the plant and protects against target pests through contact and stomach activity.

We are evaluating both Cruiser and Gaucho in different trials in combinations with different fungicides. The largeste study will be finalized after the 2007 growing season and data will then be available.

Chapter 3

Antifungal Medication

An antifungal medication is a pharmaceutical fungicide used to treat mycoses such as athlete's foot, ringworm, candidiasis (thrush), serious systemic infections such as cryptococcal meningitis, and others. Such drugs are usually obtained by a doctor's prescription or purchased over-the-counter.

Adverse Effects

Apart from side effects like liver damage or affecting estrogen levels, many antifungal medicines can cause allergic reactions in people. For example, the azole group of drugs is known to have caused anaphylaxis.

There are also many drug interactions. Patients must read in detail the enclosed data sheet(s) of the medicine. For example, the azole antifungals such as ketoconazole or itraconazole can be both substrates and inhibitors of the P-glycoprotein, which (among other functions) excretes toxins and drugs into the intestines.

Azole antifungals also are both substrates and inhibitors of the cytochrome P450 family CYP3A4, causing increased concentration when admini-stering, for example, calcium channel blockers, immuno-suppressants, chemotherapeutic drugs, benzo-diazepines, tricyclic antidepressants, macrolides and SSRIs.

Classes

Polyene Antifungals: A polyene is a molecule with multiple conjugated double bonds. A polyene antifungal is a macrocyclic polyene with a heavily hydroxylated region on the ring opposite the conjugated system. This makes polyene antifungals amphiphilic. The polyene antimycotics bind with sterols in the fungal cell membrane, principally ergosterol. This changes the transition temperature (Tg) of the cell membrane, thereby placing the membrane in a less fluid, more crystalline state. (In ordinary circumstances membrane sterols reduce the packing of the phospholipid bilayer making the plasma membrane more fluid.)

As a result, the cell's contents including monovalent ions (K+, Na+, H+, and Cl-), small organic molecules leak and this is regarded one of the primary ways cell dies. Animal cells contain cholesterol instead of ergosterol and so they are much less susceptible.

However, at therapeutic doses, some amphotericin B may bind to animal membrane cholesterol, increasing the risk of human toxicity. Amphotericin B is neph-rotoxic when given intravenously. As a polyene's hydrophobic chain is shortened, its sterol binding activity is increased. Therefore, further reduction of the hydrophobic chain may result in it binding to cholesterol, making it toxic to animals.

- Amphotericin B
- Candicidin
- Filipin – 35 carbons, binds to cholesterol (toxic)
- Hamycin
- Natamycin – 33 carbons, binds well to ergosterol
- Nystatin
- Rimocidin

Imidazole, Triazole, and Thiazole Antifungals

Azole antifungal drugs inhibit the enzyme lanosterol 14 á-demethylase; the enzyme necessary to convert lanosterol to ergosterol. Depletion of ergosterol in fungal membrane disrupts the structure and many functions of fungal membrane leading to inhibition of fungal growth.

Imidazoles

- Bifonazole
- Butoconazole
- Clotrimazole

- Econazole
- Fenticonazole
- Isoconazole
- Ketoconazole
- Miconazole
- Omoconazole
- Oxiconazole
- Sertaconazole
- Sulconazole
- Tioconazole

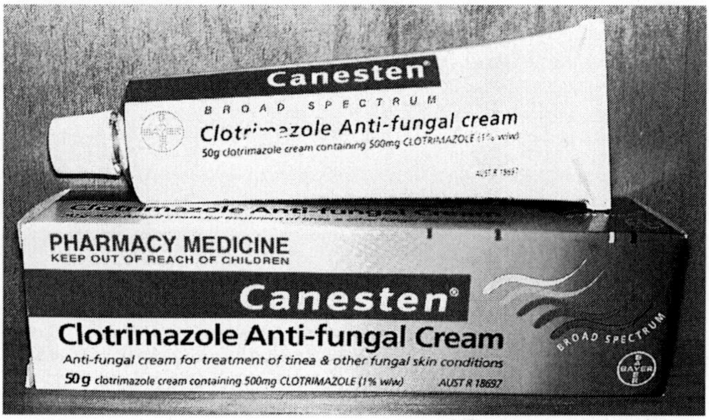

Figure: *Canesten (clotrimazole) antifungal cream*

Triazoles

- Albaconazole
- Fluconazole
- Isavuconazole
- Itraconazole
- Posaconazole
- Ravuconazole
- Terconazole
- Voriconazole

Thiazoles

- Abafungin

Allylamines

Allylamines inhibit squalene epoxidase, another enzyme required for ergosterol synthesis:

- Amorolfin
- Butenafine
- Naftifine
- Terbinafine

Echinocandins

Echinocandins may be used for systemic fungal infections in immunocompromised patients, they inhibit the synthesis of glucan in the cell wall via the enzyme 1,3-â glucan synthase:

- Anidulafungin
- Caspofungin
- Micafungin

Echinocandins are poorly absorbed when administered orally. When administered by injection they will reach most tissues and organs with concentrations sufficient to treat localized and systemic fungal infections.

Others

- *Benzoic acid* – has antifugal properties, but must be combined with a keratolytic agent such as in Whitfield's ointment
- *Ciclopirox* – (ciclopirox olamine) – is a hydroxypyridone antifungal which interferes with active membrane transport, cell membrane integrity, and fungal respiratory processes. It is most useful against tinea versicolour.
- *Flucytosine or 5-fluorocytosine* – an antimetabolite pyrimidine analog
- *Griseofulvin* – binds to polymerized microtubules and inhibits fungal mitosis
- *Haloprogin* – discontinued due to the emergence of more modern antifungals with fewer side effects
- *Polygodial* – strong and fast-acting *in-vitro* antifungal activity against *Candida albicans*.

- *Tolnaftate* – a thiocarbamate antifungal, which inhibits fungal squalene epoxidase (similar mechanism to allylamines like terbinafine)
- *Undecylenic acid* – an unsaturated fatty acid derived from natural castor oil; fungistatic, antibacterial, antiviral, and inhibits *Candida morphogenesis*
- *Crystal violet* – a triarylmethane dye, it has antibacterial, antifungal, and anthelmintic properties and was formerly important as a topical antiseptic.

Alternatives

Research conducted in 1996 indicated the following substances or essential oils had antifungal properties:

- *Oregano* – the most powerful anti-fungal of the essential oils, and possess significant activity against Candida Albicans. The minimum inhibitory concentration against C. albicans has been found to be <0.1ug per ml. In contrast, Caprylic Acid (a mixture of calcium and magnesium salts, a natural anti-fungal fatty acid), is 0.5ug.
- *Allicin* – created from crushing garlic
- *Citronella oil* – obtained from the leaves and stems of different species of *Cymbopogon* (lemon grass)
- *Coconut oil* – medium-chain triglycerides in the oil have antifungal activities
- *Iodine* – Lugol's iodine
- Lemon myrtle
- Neem seed oil
- Olive leaf
- Orange oil
- Palmarosa oil
- Patchouli
- *Selenium* – in dietary supplements or natural food sources, particularly Brazil nuts
- *Tea tree oil* – ISO 4730 ("oil of melaleuca, terpinen-4-ol type")

- *Zinc* – in dietary supplements or natural food sources, including pumpkin seeds and chickpeas
- Horopito (*Pseudowintera colourata*) leaf contains the antifungal compound polygodial.

Researchers at Tel Aviv University's Department of Plant Sciences published a study in 2009 indicating carnivorous plants, such as the Venus flytrap, contain compounds that may be useful in providing a new class of antifungal drugs for use in humans, for fungal infections that are resistant to current drugs.

Mechanism of Action

Antifungals work by exploiting differences between mammalian and fungal cells to kill the fungal organism without dangerous effects on the host. Unlike bacteria, both fungi and humans are eukaryotes. Thus, fungal and human cells are similar at the molecular level. This makes it more difficult to find or design drugs that target fungi without affecting human cells. As a consequence, many antifungal drugs cause side effects. Some of these side effects can be life-threatening if the drugs are not used properly.

Antidandruff Shampoos

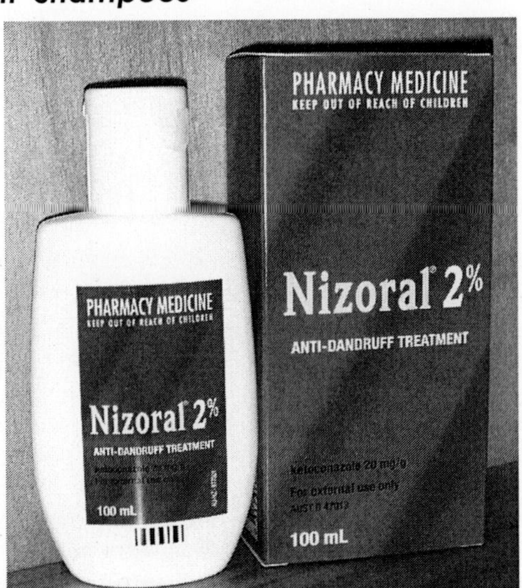

Figure: *Nizoral (ketoconazole) 2% shampoo*

Antifungal agents (such as ketoconazole) are often found in antidandruff shampoos. The antifungal drugs inhibit the yeast *Malassezia globosa* which encourages seborrhoeic dermatitis and tinea versicolour.

Agent	*Trade Names*	*Medical Applications*
Ketoconazole	Nizoral, Fungoral and Sebizole	Preliminary findings, research and studies including the completion of a small controlled clinical trial have produced data suggesting ketoconazole shampoo is effective as a hair loss treatment in men with androgenic alopecia. Larger controlled clinical studies are still needed to evaluate the ideal dosage, formulation, and to determine the routine of treatment for this condition, thus ketoconazole shampoo is not FDA-approved for this indication.
Ciclopirox olamine	Loprox	The cream and lotion form of this agent is used to treat fungal infections of the skin. The lacquer form is used as part of a treatment plan to treat fungal infections of the nails. The shampoo form is used to treat and prevent dandruff or to treat seborrhoeic dermatitis.
Piroctone olamine	Octopirox and Nivea Complete Control	Piroctone olamine is sometimes used as an antifungal agent, and it often used in dandruff shampoos in lieu of zinc. Piroctone olamine is said to be less toxic than other antidandruff agents, often bypassing some of the normal FDA warnings, but still must be used with care, and only externally.

Contd.

Contd.

Zinc pyrithione	Head & Shoulders, Johnson and Johnson ZP-11, Clinic All Clear, Pantene Pro V and Sikkai Powder	An antifungal and antibacterial Anagent first reported in the 1930s, zinc pyrithione is best known for its use in the treatment of dandruff and seborrhoeic dermatitis. It also has antibacterial properties and is effective against many pathogens from the *Streptococcus* and *Staphylococcus* genera. Its other medical appli-cations include treatments of psoriasis, eczema, ringworm, fungus, athletes foot, dry skin, atopic dermatitis, tinea, and vitiligo.
Selenium sulfide	Selsun Blue, Head & Shoulders and Vichy Dercos Anti Dandruff Shampoo	Selenium sulfide is available as a 1% and 2.5% lotion and shampoo. In some countries, the higher-strength preparations require a doctor's prescription. The shampoo is used to treat dandruff and seborrhea of the scalp, and the lotion is used to treat tinea versicolour, a fungal infection of the skin.
Tar	Neutrogena T-Gel	Is effective as a therapeutic treatment to control scalp itching and flaking symptomatic of scalp psoriasis, eczema, seborrhoeic dermatitis and dandruff.
Tea tree oil	Dr. Bronner's Castile Soap	It is used topically as an ingredient in creams, ointments, lotions, soaps, and shampoos. In addition to antifungal properties, tea tree oil has antiseptic, antibacterial, and antiviral actions. It is also effective against mites (such as scabies), and lice (such as head lice).

Azoxystrobin

Azoxystrobin (brand name Amistar, Syngenta) is a fungicide commonly used in agriculture. The substance is used as an active agent protecting plants and fruit/vegetables from fungal diseases.

Origin

Azoxystrobin was discovered during research on *Oudemansiella mucida* and *Strobilurus tenacellus*, which are small white or brown coloured mushrooms commonly found in the Czech forests.

Not bigger than a few centimetres, these mushrooms attracted attention of scientists because of their remarkable ability to defend themselves.

Their defense mechanism is based on the secretion of two substances, strobilurin A and oudemansin A. These substances allow them to keep their competitors at a distance and even destroy them when in range. Observations of this mechanism led directly to research that resulted in azoxystrobin.

Activity

After synthesizing experimental analogs of both substances (over 1400 were tested), azoxystrobin was found to be the most active and stable combination. The pharmacophore of azoxystrobine is the â-methoxyacrylate portion, which is present in the active compounds from both *Oudemansiella mucida* and *Strobilurus tenacellus*. Azoxystrobin binds very tightly to the Qo site of Complex III of the mitochondrial electron transport chain, thereby ultimately preventing the generation of ATP.

Efficacy

Azoxystrobin possesses the broadest spectrum of activity of all presently known antifungals. It is presently the only counteragent that has the ability to protect against the 4 big groups of fungal diseases:

- *Ascomycota:* Septoria
- *Deuteromycota:* Pyricularia (rice harvesting)
- *Basidiomycota:* Stripe rust
- *Oomycota:* Water mould (grape harvesting)

Examples

Azoxystrobin is widely used in farming, particularly in wheat farming. Applying agents containing azoxystrobin provides protection against many types of diseases, including:

- Wheat septoria
- Septoria leaf spot
- Wheat leaf rust (Puccinia recondita)
- Rye leafrust (Puccinia triticina)
- Powdery mildew
- Downy mildew
- Stripe rust
- Haustorium
- Pyrenophora teres

Practical Use

- Grain farming
- Banana transport
- grapes, both table & wine

Ecotoxicology

Azoxystrobin has a favourable ecotoxicological profile, meeting the expectations of agricultural demand.

Azoxystrobin is broken down into the soil. Its toxicity is low for mammals, birds, bees, insects, and earthworms. "Azoxystrobin is classified as very toxic to aquatic organisms and [its main degradation product] R234886 as very harmful. [A recent] study shows that azoxystrobin and R234886 can leach through loamy soils for a long period of time following application of the pesticide and thereby pose a potential threat to vulnerable aquatic environments and drinking water resources."

Food Residues

In recent surveillance by the New Zealand Ministry for Primary Industries, 6 out of 24 olive oil samples contained residues of azoxystrobin or the fungicide propiconazole in excess of the maximum residue limit (MRL) set for agricultural chemicals. "When the processing factor is taken into account, these results indicate that the raw olives would have likely breached the MRLs for those compounds." In the Ministry's judgement, this violation did not pose health or food safety concerns.

Barium Borate

Barium borate is an inorganic compound, a borate of barium with a chemical formula BaB_2O_4 or $Ba(BO_2)_2$. It is available as a hydrate or dehydrated form, as white powder or colourless crystals. The crystals exist in the high-temperature á phase and low-temperature â phase, abbreviated as BBO; both phases are birefringent, and BBO is a common nonlinear optical material.

Properties

Barium borate exists in two major crystalline forms, alpha and beta; the low-temperature beta phase converts into the alpha phase upon heating to 925 °C. â-Barium borate (BBO) differs from the á form by the positions of the barium ions within the crystal. Both phases are birefringent, however the á phase possesses centric symmetry and thus does not have the same nonlinear properties as the â phase.

Alpha barium borate, á-BaB_2O_4 is an optical material with a very wide optical transmission window from about 190 nm to 3500 nm. It has good mechanical properties and is a suitable material for high-power ultraviolet polarization optics. It can replace calcite, titanium dioxide or lithium niobate in Glan-Taylor prisms, Glan-Thompson prisms, walk-off beam splitters and other optical components.

It has low hygroscopicity and its Mohs hardness is 4.5. Its damage threshold is 1 GW/cm^2 at 1064 nm and 500 MW/cm^2 at 355 nm. Beta barium borate, â-BaB_2O_4, is a nonlinear optical material which is transparent in the range ~190–3300 nm. Its Mohs hardness is also 4.5.

Barium borate has strong negative uniaxial birefringence and can be phase-matched for type I (*ooe*) second harmonic generation from 409.6 to 3500 nm. The temperature sensitivity of the indices of refraction is low, leading to an unusually large (55 °C) temperature phase-matching bandwidth.

Synthesis

Barium borate can be prepared by reaction of an aqueous solution of boric acid with barium hydroxide. The prepared ã-barium borate contains water of crystallization that can not be completely removed by drying at 120 °C. Dehydrated ã-barium borate can be prepared by heating to 300–400 °C. Calcination at about 600–800 °C causes complete conversion to the â form. BBO prepared by this method does not contain trace amounts of BaB_2O_2

BBO crystals for nonlinear optics can be grown from fluxed melt of barium borate, sodium oxide and sodium chloride.

Thin films of barium borate can be prepared by MOCVD from barium(II) hydro-tri(1-pyrazolyl)borate. Different phases can be obtained depending on deposition temperatures. Thin films of beta-barium borate can be prepared by sol-gel synthesis.

Barium borate monohydrate is prepared from the solution of barium sulfide and sodium tetraborate. It is a white powder. It is used as an additive to e.g. paints as flame retardant, mold inhibitor, and corrosion inhibitor. It is also used as a white pigment.

Barium borate dihydrate is prepared from the solution of sodium metaborate and barium chloride at 90–95 °C. After cooling to room temperature, white powder is precipitated. Barium borate dihydrate loses water at above 140 °C. It is used as a flame retardant for paints, textiles, and paper.

Applications

BBO is a popular nonlinear optical crystal. Barium borate is a bactericide and fungicide. It is added to paints, coatings, adhesives, plastics, and paper and paper products.

Barium borate is resistant to moisture and ultraviolet radiation. It can act as UV stabilizer for polyvinyl chloride.

The solubility of barium borate is a disadvantage when used as a pigment. Silica-coated powders are available. The alkaline properties and the anodic passivation properties of the borate ion enhance the anticorrosion performance. Commonly available barium metaborate pigment comes in three grades; Grade I is a barium metaborate itself, grade II is compounded with 27% zinc oxide, and grade III is compounded with 18%

of zinc oxide and 29% calcium sulfate. Barium borate shows synergistic performance with zinc borate. Barium borate is used as a flux in some barium titanate and lead zirconate EIA Class 2 dielectric ceramic formulations for ceramic capacitors, in amount of about 2%. The barium-boron ratio is critical for flux performance; BaB_2O_2 content adversely affects the performance of the flux. Barium borate-fly ash glass can be used as radiation shielding. Such glasses are superior in performance to concrete and to other barium borate glasses.

Benomyl

Benomyl (also marketed as Benlate) is a fungicide introduced in 1968 by DuPont. It is a systemic benzimidazole fungicide that is selectively toxic to microorganisms and invertebrates, especially earthworms. Benomyl binds to microtubules, interfering with cell functions, such as meiosis and intracellular transportation.

The selective toxicity of benomyl as a fungicide is possibly due to its heightened effect on fungal rather than mammalian microtubules. Due to the development and worldwide prevalence of resistance of parasitic fungi to benomyl, it and similar pesticide compounds became largely ineffective. High legal costs associated with it caused DuPont to cease its production in 2001 after 33 years on the market, and voluntarily requested cancellation for its registration. However, as DuPont's patents expired long ago and in some countries benomyl's registration has not been revoked, other manufacturers still produce it.

Toxicity

Benomyl is of such a low toxicity to mammals, it has been impossible to administer doses large enough to establish an LD_{50}. It has an arbitrary LD_{50} of "greater than 10,000 mg/kg/day for rats". Skin irritation may occur through industrial exposure, and florists, mushroom pickers and floriculturists have reported allergic reactions to benomyl.

In a laboratory study, dogs fed benomyl in their diets for three months developed no major toxic effects, but did show evidence of altered liver function at the highest dose (150 mg/kg). With longer exposure, more severe liver damage occurred, including cirrhosis.

The US Environmental Protection Agency classified benomyl as a possible carcinogen. Carcinogenic studies have produced conflicting results. A two-year experimental mouse study has shown it "probably" causes an increase in liver tumours. The British Ministry of Agriculture Fisheries and Food took the view this was brought about by the hepatotoxic effect of benomyl. In 1993, the Observer, a UK national newspaper, published a series of articles alleging a possible link between exposure of pregnant mothers to benomyl and their children being born without eyes (anophthalmia) or with related syndromes, including reduced eyes and blindness, due to severe damage of the optic stem. The newspaper cited a number of suspected clusters in the UK that may have corresponded to areas of benomyl use. Studies have shown eye defects can occur at relatively high doses. A test in which rats were dosed orally demonstrated evidence of microphthalmia at dose levels of 62.5 mg/kg and above.

Birth Defects

In 1996, a Miami jury awarded US$ 4 million to a child whose mother was exposed in pregnancy to Benlate. The child was born without eyes. The mother had been exposed to an unusually high dose of Benlate through her occupation, during pregnancy. An important issue in the case was whether the timing of exposure - during the formation of the optic nerve in the foetus - was critical, as well as the magnitude of exposure. The case was prosecuted by the Ferraro Law Firm.

In October 2008, DuPont paid confidential settlements to two New Zealand families whose children were born with either anophthalmia or other birth defects. The mother of one of the children had been exposed to Benlate while working as a Christchurch parks worker before his birth.

A Benlate compensation case involving an English boy from Essex born without eyes is also due to be heard shortly in the US.

Environmental Effects

Benomyl binds strongly to soil and does not dissolve in water to any great extent. It has a half-life in turf of three to six months, and in bare soil, a half-life of six months to one year.

In 1991, DuPont issued a recall of its Benlate 50DF formula due to suspected contamination with the herbicide atrazine. In the wake of the recall, many US growers blamed Benlate 50DF for destroying millions of dollars worth of crops. Growers filed over 1900 damage claims against DuPont, mostly involving ornamental crops in Florida. Subsequent testing by DuPont determined the recalled product was not contaminated with atrazine. The reason for the alleged crop damage is unclear. The Florida Department of Agriculture and Consumer Services suggested Benlate was contaminated with dibutylurea and sulfonylurea herbicides. After several years of legal argument, DuPont paid out about US$750 million in damages and out-of-court settlements. By 1993, a coalition of farm worker and environmental groups came together to form "Benlate Victims Against DuPont", a group which called for a nation-wide boycott of DuPont products.

After carrying out tests, DuPont denied Benlate was contaminated with dibutylurea and sulfonylureas and stopped compensation pay-outs. In 1995, a Florida judge rejected a complaint from the Florida Department of Agriculture that had alleged such a link.

Cellular Biology

Benomyl is used in molecular biology to study the cell cycle in yeast; in fact, the name of the protein class "Bub" (Bub1, etc.) comes from their mutant in which budding was uninhibited by benomyl.

Binapacryl

Binapacryl is an miticide and fungicide. Chemically, it is an ester derivative of dinoseb. Although binapacryl has low toxicity itself, it is readily metabolized to form dinoseb, which is toxic. International trade in binapacryl is regulated by the Rotterdam Convention.

Bordeaux Mixture

Bordeaux mixture (also called *Bordo Mix*) is a mixture of copper(II) sulfate ($CuSO_4$) and slaked lime used as a fungicide in vineyards. It is used mainly to control garden, vineyard, nursery and farm infestations of fungi, primarily downy mildew, which can result from infections of *Plasmopara viticola*. It was invented in the Bordeaux region of France, where it is known locally as *Bouillie Bordelaise*. This fungicide has been used for over a century and is still used, although the copper can leach out and pollute streams.

Main Uses

In addition to its use to control fungal infection on grape vines, the mixture is also widely used to control potato blight, peach leaf curl and apple scab. It is approved for organic use, so is often used by organic gardeners where nonorganic gardeners would prefer other controls.

History

In the 19th century, several outbreaks of vine diseases occurred among the *Vitis vinifera* vines of the classicial European wine regions. These outbreaks were caused by pests to which these vines lacked resistance, carried on vines brought to Europe as botanical specimens of American origin. These pests included not only the Great French Wine Blight caused by the aphid *Phylloxera vastatrix*, but also mildew and other diseases caused by fungi.

After the downy mildew had struck, botany professor Pierre-Marie-Alexis Millardet of the University of Bordeaux studied the disease in vineyards of the Bordeaux region. Millardet then noted that vines closest to the roads did not show mildew, while all other vines were affected. After inquiries, he found out those vines had been sprayed with a mixture of $CuSO_4$ and lime to deter passersby from eating the grapes, since this treatment was both visible and bitter-tasting. This led Millardet to conduct trials with this treatment. The trials primarily took place in the vineyards of Château Dauzac, where he was assisted by Ernest David, Dauzac's technical director. Millardet published his findings in 1885, and recommended the mixture to combat downy mildew.

In France, the use of Bordeaux mixture has also been known as the Millardet-David treatment.

Mode of Action

Bordeaux mixture achieves its effect by means of the copper ions (Cu^{2+}) of the mixture. These ions affect enzymes in the fungal spores in such a way as to prevent germination. This means Bordeaux mixture must be used pre-emptively, before the fungal disease has struck.

Preparation

Bordeaux mixture can be prepared using differing proportions of the components. In preparing it, the $CuSO_4$ and the lime are dissolved separately in water and then mixed. Calcium oxide (burnt lime) and calcium hydroxide (hydrated lime) give the same end result, since an excess of water is used in the preparation.

The conventional method of describing the mixture's composition is to give the weight of $CuSO_4$, the weight of hydrated lime and the volume of water, in that order. The percentage of the weight of $CuSO_4$ to the weight of water employed determines the concentration of the mixture. Thus a 1% Bordeaux mixture, which is typical, would have the formula 1:1:100, with the first "1" representing 1 kg $CuSO_4$ (pentahydrated), the second representing 1 kg hydrated lime, and the 100 representing 100 litres (100 kg) water. As $CuSO_4$ contains 25% copper, the copper content of a 1% Bordeaux mixture would be 0.25%. The quantity of lime used can be lower than that of the $CuSO_4$. One kg of $CuSO_4$ actually requires only 0.225 kg of chemically pure hydrated lime to precipitate all the copper. Good proprietary brands of hydrated lime are now freely available, but, as even these deteriorate on storage (by absorbing carbon dioxide from the air), a ratio of less than 2:1 is seldom used, which corresponds to a 1:0.5:100 mixture.

Risks

Bordeaux mixture has been found to be harmful to fish, livestock and—due to potential build up of copper in the soil—earthworms. The chemical started to be used by the United Fruit Company throughout Latin America around 1922. The mixture was nicknamed *perico*, or "parakeet", because it would turn workers completely blue. Many workers would get sick or die of poisoning due to the toxic chemical.

Bromine Monochloride

Bromine monochloride, also called bromine(I) chloride, bromochloride, and bromine chloride, is an interhalogen inorganic compound with chemical formula BrCl. It is an unstable brownish yellow gas with boiling point 5°C and melting point "66 °C. Its CAS number is 13863-41-7 and its EINECS number is 237-601-4. It is a strong oxidizing agent.

Uses

Bromine monochloride is used in analytical chemistry in determining low levels of mercury, to quantitatively oxidize mercury in the sample to Hg(II) state. Bromine monochloride is used as a biocide, specifically as an algaecide, fungicide, and disinfectant, in industrial recirculating cooling water systems. Addition of bromine monochloride is used in some types of Li-SO_2 batteries to increase voltage and energy density.

Burgundy Mixture

Burgundy mixture, named after the French district where it was first used to treat grapes and vines, is a mixture of copper sulfate and sodium carbonate. This mixture, which can have an overall copper concentration within the range of 1% through 20%, is used as a fungicidal spray for trees and small fruits.

History

Similar to the Bordeaux Mixture, one of the earliest fungicides in use, Burgundy Mixture, also known as "Sal soda Bordeaux," is used as a fungus preventative applicant on plants before fungi have appeared. Bordeaux Mixture contains copper sulfate, $CuSO_4$, and hydrated lime, $Ca(OH)_2$, while Burgundy Mixture contains copper sulfate, $CuSO_4$, and sodium carbonate, Na_2CO_3. First used around 1885, Burgundy Mixture has since been replaced by synthetic organic compounds, or by compounds that contain copper in a non-reactive, chelated form. This helps to prevent the accumulation of high levels of copper in sediments surrounding the plants.

Synthesis and Composition

Burgundy Mixture is made by combining dissolved copper sulfate and dissolved sodium carbonate. Dissolved copper sulfate ratios generally range from 1:1 to 1:18. Sodium carbonate is generally added in higher quantities and at a dissolved ratio of 1:1.5. Over time, the sodium carbonate will crystallize out of solution, and the closer the copper sulfate to sodium carbonate mixture is to 1:1 ratios, the faster this process occurs. This property is one key factor in the general discontinued usage of Burgundy Mixture, as the mixture must be mixed and used shortly before intended utilization.

Uses and Mode of Action

Burgundy Mixture is used as a preemptive fungicide prevention for trees and small fruits. This occurs because the Cu(II) ions are capable of interfering with enzymes found within the spores of many fungi, preventing germination from occurring. Unfortunately, the mechanism for copper antifungal properties is not well understood, though it is often thought that interactions between the copper and negatively charged portions of the cellular walls of the bacteria promote an altered shape and increased membrane permeability, which alters the homeostasis of the cell and can lead to insufficient uptake and storage of essential nutrients and ions.

Captafol

Captafol is a fungicide. It is used to control almost all fungal diseases of plants except powdery mildews. It is believed to be a human carcinogen, and production for use as a fungicide in the United States stopped in 1987. Its continued use from existing stocks was allowed, but in 1999 the Environmental Protection Agency banned its use on all crops except onions, potatoes, and tomatoes. In 2006 even these exceptions were disallowed, so currently its use on all crops is banned in the United States. Several other countries have followed suit since 2000, and as of 2010, no countries are known to allow the use of captafol on food crops. Captafol was disclosed in US patent 3,178,447 (1965). Its synergistic mixture with thiabendazol was described in US patent 4092422 (1978). International trade in captafol is regulated by the Rotterdam Convention.

Captan

Captan is the name of a general use pesticide (GUP) that belongs to the phthalimide class of fungicides. Though it can be applied on its own, Captan is often added as a component of other pesticide mixtures. It is used to control diseases on a number of fruits and vegetables as well as ornamental plants. It also improves the outward appearance of many fruits, making them brighter and healthier-looking. Captan is utilized by both home and agricultural growers and is often applied during apple production. Captan is cited as Group B2, a probable human carcinogen by the EPA.

Carbendazim

Carbendazim is a widely used, broad-spectrum benzimidazole fungicide and a metabolite of benomyl. It is also employed as a casting worm control agent in amenity turf situations such as golf greens, tennis courts etc. and in some countries is licensed for that use only. The fungicide is used to control plant diseases in cereals and fruits, including citrus, bananas, strawberries, pineapples, and pomes. It is also controversially used in Queensland, Australia on macadamia plantations. A 4.7% solution of carbendazim hydrochloride, sold as Eertavas, is marketed as a treatment for Dutch elm disease.

Studies have found high doses of carbendazim cause infertility and destroy the testicles of laboratory animals. Maximum pesticide residue limits (MRLs) have reduced since discovering its harmful effects. The MRLs for fresh produce in the EU are now between 0.1 and 0.7 mg/kg with the exception of loquat, which is 2 mg/kg. The limits for more commonly consumed citrus and pomme fruits are between 0.1 and 0.2 mg/kg.

Chloranil

Chloranil is a quinone with the molecular formula $C_6Cl_4O_2$. Also known as tetrachloro-1,4-benzoquinone, it is a yellow solid. Like the parent benzoquinone, chloranil is a planar molecule that functions as a mild oxidant.

Reagent

Chloroanil, moreso than benzil, serves as a hydrogen acceptor. It is used for the aromatization reactions, such as the conversion of cyclohexadienes to the benzene derivatives.

Chloranil is used to test for free secondary amines. This test is useful for checking for the presence of proline derivatives. It is also a good test for the successful deprotection of a secondary amine. Secondary amines react with p-chloranil to give a brown/red/orange derivative, the colour depending on the amine. In these reactions, the amine displaces chloride from the ring of the quinone.

Chlorothalonil

Chlorothalonil (2,4,5,6-tetrachloroisophthalonitrile) is a polychlorinated aromatic mainly used as a broad spectrum, nonsystemic fungicide, with other uses as a wood protectant, pesticide, acaricide, and to control mold, mildew, bacteria, algae. Chlorothalonil-containing products are sold under the names Bravo, Echo, and Daconil. It was first registered for use in the US in 1966. In 1997, the most recent year for which data are available, it was the third most used fungicide in the US, behind only sulfur and copper, with some 12 million lb used in agriculture alone that year. Including nonagricultural uses, the EPA estimates, on average, almost 15 million lb were used annually from 1990-1996.

Uses

In the US, chlorothalonil is used predominantly on peanuts (about 34% of usage), potatoes (about 12%), and tomatoes (about 7%), though the EPA recognizes its use on many other crops. It is also used on golf courses and lawns (about 10%) and as a preservative additive in some paints (about 13%), resins, emulsions, and coatings.

Chlorothalonil is commercially available in many different formulations and delivery methods. It is applied as a dust, dry or water-soluble grains, a wettable powder, a liquid spray, a fog, and a dip. It may be applied by hand, by ground sprayer, or by aircraft.

Mechanism of Action

Chlorothalonil reduces fungal intracellular glutathione molecules to alternate forms which cannot participate in essential enzymatic reactions, ultimately leading to cell death, similar to the mechanism of trichloromethyl sulfenyl.

Toxicity

Acute: According to the United States Environmental Protection Agency, chlorothalonil is a toxicity category I eye irritant, producing severe eye irritation. It is in toxicity category II, "moderately toxic", if inhaled (inhaled LD_{50} 0.094 mg/l in rats.) For skin contact and ingestion, chlorothalonil is rated toxicity category IV]], "practically nontoxic", meaning the oral and dermal LD_{50} is greater than 10,000 mg/kg.

Chronic

Long-term exposure to chlorothalonil resulted in kidney damage and tumors in animal tests.

Carcinogenic

Chlorothalonil is a Group B2 "probable human carcinogen", based on observations of cancers and tumors of the kidneys and forestomachs in laboratory animals fed diets containing chlorothalonil.

Environmental

Chlorothalonil is highly toxic to fish and aquatic invertebrates, but not toxic to birds. At a concentration of 164 μg/l, chlorothalonil was found to kill a species of frog within a 24-hour exposure.

Contaminants

Common chlorothalonil synthesis procedures frequently result in contamination of it with small amounts of hexachlorobenzene (HCB), which is toxic. US regulations limit HCB in commercial production to 0.05% of chlorothalonil. According to the EPA report, "post-application exposure to HCB from chlorothalonil is not expected to be a concern based on the low level of HCB in chlorothalonil."

Environmental Contamination

Chlorothalonil has been detected in ambient air in Minnesota and Prince Edward Island, as well as in groundwater in Long Island, New York and Florida. In the first three cases, the contamination is presumed to have come from potato farms. Research has been done by the University of South Florida and Biologists Taegan McMahon and Jason Rohr report that chlorothalonil killed amphibians, snails, zooplankton, algae, and aquatic plants below estimated environmental concentrations previously deemed safe by the U.S. Environmental Protection Agency. The loss of these herbivores and plants freed the algae from predation and competition, which eventually resulted in algal blooms that were similar to the effects of eutrophication.

The main breakdown product of chlorothalonil is 4-hydroxy-2,5,6-trichloroisophthalonitrile (SDS-3701). It has been shown to be 30 times more acutely toxic than chlorothalonil and more persistent in the environment. Laboratory experiments have shown it can thin the eggshells of birds, but no evidence supports this happening in the environment.

Chemistry

Chlorothalonil can be produced by the direct chlorination of isophthalonitrile or by treating tetrachloroisophthaloyl amide with phosphoryl chloride. It breaks down under basic conditions, but is stable in neutral and acidic media. Technical chlorothalonil contains traces of dioxins and hexachlorobenzene, a persistent organic pollutant banned under the Stockholm Convention.

Chromated Copper Arsenate

Chromated copper arsenate (CCA) is a wood preservative that has been used for timber treatment since the mid-1930s. It is a mix of chromium, copper and arsenic (as Copper(II) arsenate) formulated as oxides or salts, and is recognizable for the greenish tint it imparts to timber. CCA was invented in 1933 by Dr. Sonti Kamesam, an Indian scientist, and was awarded its first patent (British) in 1934. CCA is known by many trade names including the worldwide brands "Ascu", "Tanalith", "Wolman", "Permacure" and "Celcure".

Mechanism of Action

The chromium acts as a chemical fixing agent and has little or no preserving properties; it helps the other chemicals to fix in the timber, binding them through chemical complexes to the wood's cellulose and lignin. The copper acts primarily to protect the wood against decay fungi and bacteria, while the arsenic is the main insecticidal component of CCA, providing protection from wood attacking insects including termites and marine borers. It also improves the weather-resistance of treated timber and may assist paint adherence in the long term.

Releases to the Environment

Over time small amounts of the CCA constituents, mainly the arsenic, may leach out of the treated timber. This is particularly the case in acidic environments. The chemicals may leach from the wood into surrounding soil, resulting in concentrations higher than naturally occurring background levels. A study found that during a 12-month period 12–13 percent of the CCA leached from treated wood buried in compost. On the other hand there have been many other studies in less aggressive soil types that show leaching to be as low as 0.5 ppm (red pine poles in service,) or up to 14 ppm (treated pine in garden beds). Soil contamination due to the presence of CCA-treated wood after 45 years is minimal.

Should any chemicals leach from the wood they are likely to bind to soil particles, especially in soils with clay or soils that are more alkaline than neutral. Although widespread restrictions followed the publication of studies which showed low-level leaching from in-situ timbers (such as children's playground equipment) into surrounding soil, a more serious risk is presented if CCA-treated timber is burnt in confined spaces such as a domestic fire or barbecue. Scrap CCA construction timber continues to be widely burnt through ignorance, in both commercial, and domestic fires.

Notwithstanding this, disposal by burning, e.g. in approved incinerators, is an acceptable option, and some energy may be captured in the process.

Limitation of Human Exposure

A number of countries have reviewed CCA during recent years and have looked at limiting the public exposure to CCA-treated timber by restricting its application in residential situations. These reviews have resulted from increasing public pressures and the possibility of CCA-treated timber posing a health hazard.

In response to these pressures the preservation industry in the USA and Canada volunteered not to use CCA for the treatment of residential timber, and on 31 December 2003 the production of CCA-treated wood for such applications became a violation of the manufacturers' labels approved by the United States Environmental Protection Agency (EPA). Exceptions were allowed, including the treatment of shakes and shingles, permanent wood foundations, and certain commercial applications.

The regulatory agencies advised however that CCA-treated timber products already in use pose no significant threat to health. Indeed CCA will continue to be used in North America in a wide variety of commercial and industrial applications such as poles, piling, retaining structures and many others.

Following the USA and Canada actions in restricting CCA, similar actions have been taken in other parts of the world, including the EU and Australia. In New Zealand the Environmental Risk Management Authority, reviewing the same data that prompted the actions elsewhere, concluded that there was no reason to restrict CCA use for any applications, but notes that few well-designed studies have been carried out of those using CCA or CCA-treated timber.

CCA timber is still in widespread use in many countries and remains an economical option for conferring durability to perishable timbers such as plantation-grown pine.

Disposal of large quantities of CCA-treated wastes or spent timber at the end of its lifecycle has been traditionally through controlled landfill sites. Such sites are lined to make them impervious in order to prevent losses to the water table and they are covered to prevent rainfall washing out any contained potential toxicants. These controlled sites handle a range of waste materials potentially more noxious than that posed by CCA-treated timber, e.g. paint-stuffs, car batteries, etc. Today, landfill sites are becoming scarcer and disposal of waste materials is becoming economically unattractive. The wood preservation and timber industries are therefore researching better ways of dealing with waste treated timber, including CCA-treated material.

Alternatives

Alternative heavy-duty preservatives include creosote and pentachlorophenol. Similar water-borne preservatives include alkaline copper quaternary compounds (ACQ), copper azole (CuAz), ammoniacal copper zinc arsenate (ACZA), copper citrate, and copper HDO (CuHDO).

Copper(I) Cyanide

Copper(I) cyanide is an inorganic compound with the formula CuCN. This off-white solid occurs in two polymorphs; impure samples can be green due to the presence of Cu(II) impurities. The compound is useful as a catalyst, in electroplating copper, and as a reagent in the preparation of nitriles.

Structure

Copper cyanide is a coordination polymer. It exists in two polymorphs both of which contain -[Cu-CN]- chains made from linear copper(I) centres linked by cyanide bridges. In the high-temperature polymorph, HT-CuCN, which is isostructural with AgCN, the linear chains pack on a hexagonal lattice and adjacent chains are off set by +/- 1/3 *c*. In the low-temperature polymorph, LT-CuCN, the chains deviate from linearity and pack into rippled layers which pack in an AB fashion with chains in adjacent layers rotated by 49 °.

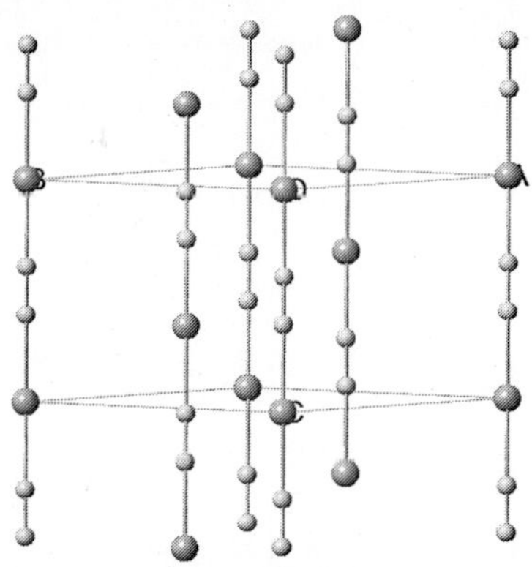

Figure: *The structure of HT-CuCN showing the chains running along the c axis. Key: copper = orange and cyan = head-to-tail disordered cyanide groups.*

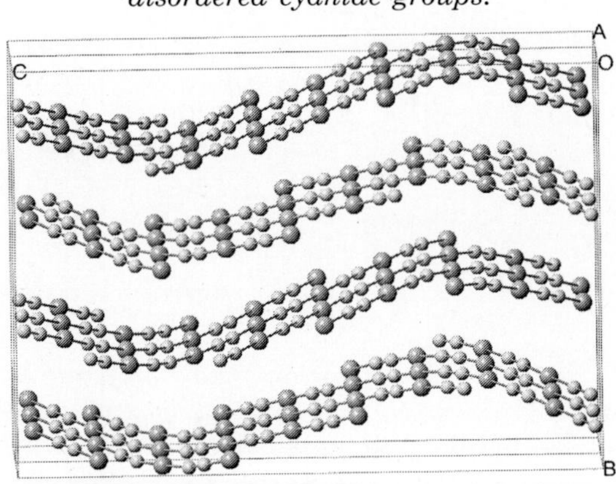

Figure: *The structure of LT-CuCN showing sheets of chains staking in an ABAB fashion. Key copper = orange and cyan = head-to-tail disordered cyanide groups.*

LT-CuCN can be converted to HT-CuCN by heating to 563 K in an inert atmosphere.

In both polymorphs the copper to carbon and copper to nitrogen bond lengths are ~1.85 Å and bridging cyanide groups show head-to-tail disorder.

Preparation

Cuprous cyanide is commercially available and is supplied as the low-temperature polymorph. It can be prepared by the reduction of copper(II) sulfate with sodium bisulfite at 60 °C, followed by the addition of sodium cyanide to precipitate pure LT-CuCN as a pale yellow powder.

$$2\ CuSO_4 + NaHSO_3 + H_2O + 2\ NaCN \rightarrow 2\ CuCN + 3\ NaHSO_4$$

On addition of sodium bisulfite the copper sulfate solution turns from blue to green, at which point the sodium cyanide is added.

The reaction is performed under mildly acidic conditions. Copper cyanide has historically been prepared by treating copper(II) sulfate with sodium cyanide, in this redox reaction, copper(I) cyanide forms together with cyanogen:

$$2\ CuSO_4 + 4\ NaCN \rightarrow 2\ CuCN + (CN)_2 + 2\ Na_2SO_4$$

Because this synthetic route produces cyanogen, uses two equivalents of sodium cyanide per equivalent of CuCN made and the resulting copper cyanide is impure it is not the industrial production method. The similarity of this reaction to that between copper sulfate and sodium iodide to form copper(I) iodide is one example of cyanide ions acting as a pseudo halide. It also explains why copper(II) cyanide, $Cu(CN)_2$, has not been synthesised.

Reactions

Copper cyanide is insoluble in water but rapidly dissolves in solutions containing CN^- to form $[Cu(CN)_3]^{2-}$ and $[Cu(CN)_4]^{3-}$, these ions are trigonal planar and tetrahedral respectively. This is in contrast to both silver and gold cyanides which form $[M(CN)_2]^-$ ions in solution. The coordination polymer $KCu(CN)_2$ contains $[Cu(CN)_2]^-$ units which link together forming helical anionic chains. Copper cyanide is also soluble in 0.88 aqueous ammonia, pyridine and N-methylpyrrolidone.

Applications

Cuprous cyanide is used for electroplating copper. The compound is useful reagent in organic synthesis, for example in the regioselective and stereoselective allylation and conjugate addition of N-Boc-2-lithiopyrrolidine and N-Boc-2-lithiopiperidine, or the copper cyanide catalyzed palladium coupling of á-lithio amines and aryl iodides.

Cinnamaldehyde

Cinnamaldehyde is the organic compound that gives cinnamon its flavour and odor. This pale yellow, viscous liquid occurs naturally in the bark of cinnamon trees and other species of the genus *Cinnamomum*. The essential oil of cinnamon bark is about 90% cinnamaldehyde.

Structure and Synthesis

Cinnamaldehyde was isolated from cinnamon essential oil in 1834 by Dumas and Péligot and synthesized in the laboratory by Chiozza in 1854.

The natural product is *trans*-cinnamaldehyde. The molecule consists of a phenyl group attached to an unsaturated aldehyde. As such, the molecule can be viewed as a derivative of acrolein. Its colour is due to the ð → ð* transition: increased conjugation in comparison with acrolein shifts this band towards the visible.

Synthesis

Several methods of laboratory synthesis exist, but cinnamaldehyde is most economically obtained from the steam distillation of the oil of cinnamon bark. The compound can be prepared from related compounds such as cinnamyl alcohol, (the alcohol form of cinnamaldehyde), but the first synthesis from unrelated compounds was the aldol condensation of benzaldehyde and acetaldehyde.

Metabolism

Cinnamaldehyde occurs widely, and closely related compounds give rise to lignin. All such compounds are biosynthesized starting from phenylalanine, which undergoes conversion. Cinnamoyl-CoA reductase is an enzyme responsible for the production of cinnamoyl-CoA from cinnamaldehyde.

Applications

As a Flavourant: The most obvious application for cinnamaldehyde is as flavouring in chewing gum, ice cream, candy, and beverages; use levels range from 9 to 4900 parts per million (ppm) (that is, less than 0.5%). It is also used in some perfumes of natural, sweet, or fruity scents. Almond, apricot, butterscotch, and other aromas may partially employ the compound for their pleasant smells. Cinnamaldehyde can be used as a food adulterant; powdered beechnut husk aromatized with cinnamaldehyde can be marketed as powdered cinnamon.

As an Agrichemical

Cinnamaldehyde is also used as a fungicide. Proven effective on over 40 different crops, cinnamaldehyde is typically applied to the root systems of plants. Its low toxicity and well-known properties make it ideal for agriculture. Cinnamaldehyde is an effective insecticide, and its scent is also known to repel animals, such as cats and dogs. It has recently been recognized as a very effective insecticide for mosquito larvae. As little as 29 ppm of cinnamaldehyde kills half of *Aedes aegypti* mosquito larvae in 24 hours.

As an Antimicrobial

Another use for cinnamaldehyde is as an antimicrobial. Researchers from the University of Illinois at Chicago (who were funded by the Wm. Wrigley Jr. Company) have found that cinnamic aldehyde, when used in Big Red, prevented oral bacterial growth by more than 50%. It is especially effective against bacteria living at the back of the tongue, reducing anaerobic bacteria populations by about 43%.

As an Anticancer Agent

Recent research documents anticancer activity of cinnamaldehyde/cinnamic aldehyde observed in cell culture and animal models of the disease. Proliferation, invasion, and tumor growth were inhibited in a murine A375 model of human melanoma, though only at high doses not achievable through dietary intake.

Miscellaneous Uses

Cinnamaldehyde is also known as a corrosion inhibitor for steel and other ferrous alloys in corrosive fluids. It can be used in combination with additional components such as dispersing agents, solvents and other surfactants. Its high refractive index of 1.6220 makes it a fairly safe and useful fluid for examining gemstone rough for inclusions.

Derivatives of Cinnamaldehyde

Numerous derivatives of cinnamaldehyde are commercially useful. Dihydrocinnamyl alcohol, which occurs naturally but is produced by double hydrogenation of cinnamaldehyde, is used to confer the fragrances of hyacinth and lilac. Cinnamyl alcohol similarly occurs naturally and has the odor of lilac. Dihydrocinnamaldehyde is produced by the selective hydrogenation of the alkene subunit. á-Amyl- and á-hexylcinnamaldehyde are important commercial fragrances, but they are not prepared from cinnamaldehyde.

Toxicology

Cinnamaldehyde is used in agriculture because of its low toxicity, but it is a skin irritant.

The level of coumarin detected in 'cinnamon' extracts and powders has concerned the German Federal Institute for Risk Assessment (BfR).

Copper(II) Arsenate

Copper arsenate ($Cu_3(AsO_4)_2.4H_2O$, or $Cu_5H_2(AsO_4)_4.2H_2O$), also called copper orthoarsenate or tricopper arsenate or tricopper orthoarsenate, is a blue or bluish-green powder insoluble in water and alcohol and soluble in aqueous ammonium and dilute acids. Its CAS number is [7778-41-8] or [10103-61-4].

Uses

Copper arsenate is an insecticide used in agriculture. It is also used as a herbicide, fungicide, and a rodenticide. It is also used as a poison in slug baits. Copper arsenate can also be a misnomer for copper arsenite, especially when meant as a pigment.

Natural Occurrences

Anhydrous copper arsenate, $Cu_3(AsO_4)_2$, is found in nature as the mineral lammerite. Copper arsenate tetrahydrate, $Cu_3(AsO_4)_2.4H_2O$, occurs naturally as the mineral rollandite.

Related Compounds

Copper arsenate hydroxide or basic copper arsenate ($Cu(OH)AsO_4$) is a basic variant with CAS number [16102-92-4]. It is found naturally as the mineral olivenite. It is used as an insecticide, fungicide, and miticide. Its use is banned in Thailand since 2001.[1]

Cycloheximide

Cycloheximide is an inhibitor of protein biosynthesis in eukaryotic organisms, produced by the bacterium *Streptomyces griseus*. Cycloheximide exerts its effect by interfering with the translocation step in protein synthesis (movement of two tRNA molecules and mRNA in relation to the ribosome) thus blocking translational elongation.

Cycloheximide is widely used in biomedical research to inhibit protein synthesis in eukaryotic cells studied *in vitro* (*i.e.* outside of organisms). It is inexpensive and works rapidly. Its effects are rapidly reversed by simply removing it from the culture medium.

Due to significant toxic side effects, including DNA damage, teratogenesis, and other reproductive effects (including birth defects and toxicity to sperm), cycloheximide is generally used only in *in vitro* research applications, and is not suitable for human use as a therapeutic compound. Although it has been used as a fungicide in agricultural applications, this application is now decreasing as the health risks have become better understood.

Cycloheximide is degraded by alkali ($pH > 7$), decontamination of work surfaces and containers can be achieved by washing with a non-harmful alkali solution such as soap. Since cycloheximide is an effective inhibitor of protein biosynthesis in eukaryotes only, it may be used to distinguish between proteins translated in the mitochondria and proteins translated in the cytosol.

mRNA translated in cytosol or ER from mRNA derived from the nucleus will not be expressed in the presence of cycloheximide. Conversely, translation using mitochondrial ribosomes is unaffected by cycloheximide, and mitochondrial genes will continue to be expressed.

Experimental Applications

Cycloheximide can be used as an experimental tool in molecular biology to determine the half-life of a protein. Treating cells with cycloheximide in a time-course experiment followed by Western blotting of the cell lysates for the protein of interest can show differences in protein half-life. Cycloheximide treatment provides the ability to observe the half-life of a protein without confounding contributions from transcription or translation.

It is used as a plant growth regulator to stimulate ethylene production. It is used as a rodenticide and other animal pesticide. It is also used in media to detect unwanted bacteria by suppressing yeasts and molds in beer fermentation.

Davicil

Davicil is a chlorinated pyridine derivative with antimicrobial properties, which is used as a fungicide. It can be allergenic in humans and produce contact dermatitis.

Dehydroacetic Acid

Dehydroacetic acid is an organic compound. The compound is classified as a pyrone derivative and is used mostly as a fungicide and bactericide. The sodium salt, sodium dehydroacetate, is often used in place of dehydroacetic acid because of its greater solubility in water. It is prepared by the base-catalysed dimerization of diketene.

Applications

Industrially, it is also used as a plasticizer in a synthetic resins. It is used to reduce pickle bloating as a preservative for squash and strawberries. Also used in antienzyme toothpastes. Clinical tests show little to no evidence regarding the ingredient's toxicity and potential for irritation.

Diphenylamine

Diphenylamine is an organic compound with the formula $(C_6H_5)_2NH$ belonging to the class of aromatic amines. The molecule is "aromatic" because of the presence of a benzene ring in its structure, and an "amine" because of the presence of a basic N atom. The compound exists as a colourless solid at ordinary room temperature, but samples are often yellow due to oxidized impurities. Diphenylamine dissolves well in many common organic solvents, but is insoluble in water.

Preparation and Reactivity

Diphenylamine is manufactured by the thermal deamination of aniline over oxide catalysts:

$$2\ C_6H_5NH_2 \rightarrow (C_6H_5)_2NH + NH_3$$

It is a weak base, with a K_b of 10^{-14}. With strong acids, it forms salts (e.g. reaction with sulfuric acid gives the bisulfate $[(C_6H_5)_2NH_2]^+[HSO_4]^-$ as a white or yellowish powder with m.p. 123-125°C.

Applications

Diphenylamine is used as a pre- or postharvest scald inhibitor for apples. Its anti-scald activity is the result of its antioxidant properties, which protect the apple skin from the oxidation products of alpha-farnesene during storage.

Diphenylamine derivatives are also useful. Ring-alkylated derivatives of diphenylamine are used as "antiozinates" in the manufacture of rubber products, reflecting the antioxidant nature of aniline derivatives. The compound undergoes various cyclisaton reactions. With sulfur, it gives phenothiazine, a precursor to certain pharmaceuticals.

$$(C_6H_5)_2NH + 2\ S \rightarrow S(C_6H_4)_2NH + H_2S$$

With iodine, it cyclises to carbazole:

$$(C_6H_5)_2NH + I_2 \rightarrow (C_6H_4)_2NH + 2\ HI$$

Arylation with iodobenzene gives triphenylamine.

Diphenylamine finds niche use as a test for nitrates.

Epoxiconazole

Epoxiconazole is a fungicide active ingredient from the class of azoles developed to protect crops. In particular, the substance inhibits the metabolism of fungi cells infesting useful plants, and thereby prevents the growth of the mycelia (fungal cells). Epoxiconazole also limits the production of conidia (mitospores). Epoxiconazole was introduced to the market by BASF SE in 1993 and can be found in many products and product mixtures targeting a large number of pathogens in various crops. Crops are, for example, cereals (mainly wheat, barley, rye and triticale), soybeans, banana, rice, coffee, turnips, and red as well as sugar beets.

Use

World food supply is projected to face additional stress in the future; within the next decade cereal demands in Europe alone are expected to increase by 21%, while production is projected to increase by only 7%. At the same time, two main cereal diseases, leaf blotch (Septoria tritici) and rust (Puccinia triticina), are responsible for up to 30% yield losses. Failure to adequately control these diseases would affect both food availability and food quality.

Resistance

Certain plant pathogens develop resistance to fungicides. In contrast to the relatively rapid development of resistance to strobilurins, azole fungicides like Epoxiconazole have maintained their effectiveness controlling key wheat diseases for over two decades. According to a study conducted by the Home Grown Cereals Authority (HGCA), Epoxiconazole was one of two triazole fungicides (the other being prothioconazole) reported to still provide a high level of eradicative and protective control of Septoria tritici. Additional classes of fungicides like contact fungicides, strobilurins or carboxamides are available to farmers. In the latter case, the best activity rates are achieved in mixtures with triazoles.

Mode of Action

As an azole, Epoxiconazole, actively stops the production of new fungi spores and inhibits the biosynthesis of existing hostile cells. Epoxiconazole works as an eradicant by encapsulating fungal haustoria, which are then cut off from their nutrient supply and therefore die. Some fungicide interactions can actually lead to increased production of mycotoxins, which are normal metabolic products of fungi, and it has been found that the inclusion of triazoles, like Epoxiconazole, in the fungicide mix may be necessary to limit mycotoxin levels.

Registrative Status

The EU's pesticide directive (91/414 EEC) currently lists Epoxiconazole in Annex I. Only active ingredients listed there may be used in plant protection products and sold to farmers. Substances can only be listed in Annex I if extensive data regarding the physical chemical properties, the fate and behaviour in the environment and the toxicological properties have been evaluated by EU authorities and EFSA. Only if a substance has been proven to be safe for users, consumers and the environment when used according to instructions it is included. Annex I listings are granted for ten years. Epoxiconazole's current listing expires on 30 April 2019; European farmers can use products containing Epoxiconazole, depending on their national registration, at least until the date of expiration for Annex I listing.

Falcarindiol

Falcarindiol is a polyacetylene found in carrot roots which has antifungal activity. Falcarindiol is the main compound responsible for bitterness in carrots.

Falcarinol

Falcarinol (carotatoxin) is a natural pesticide and fatty alcohol found in carrots (*Daucus carota*), red ginseng (*Panax ginseng*) and ivy. In carrots, it occurs in a concentration of approximately 2 mg/kg. As a toxin, it protects roots from fungal diseases, such as liquorice rot that causes black spots on the roots during storage.

Chemistry

Falcarinol is a polyyne with two carbon carbon triple bonds and two double bonds. Falcarinol can cause allergic and irritant contact dermatitis. It is structurally related to the oenanthotoxin and cicutoxin.

Biological Effects

It was shown that falcarinol acts as a covalent cannabinoid receptor type 1 inverse agonist and blocks the effect of anandamide in keratinocytes, leading to pro-allergic effects in human skin. Preliminary research in animal models suggest that falcarinol may have a protective effect against certain types of cancer. Laboratory rats fed a diet containing raw carrots or isolated falcarinol were a third less likely to develop full-scale tumors induced by azoxymethane than those in a control group.

Normal consumption of carrots doesn't causes any toxic effect in humans. However, when falcarinol is delivered in high doses to laboratory animals, it causes neurotoxical problems.

Famoxadone

Famoxadone is a fungicide to protect agricultural products against various fungal diseases on fruiting vegetables, tomatoes, potatoes, curcurbits, lettuce and grapes. It is used in combination with cymoxanil.

Fenamidone

Fenamidone is a foliar fungicide used on grapes, ornamentals, potatoes, tobacco, and vegetables. It exerts its fungicidal effects by acting as a Q_o inhibitor.

Fenarimol

Fenarimol, sold under the tradenames Bloc, Rimidin and Rubigan, is a fungicide which acts against rusts, blackspot and mildew fungi.

It is used on ornamental plants, trees, lawns, tomatoes, peppers, eggplants, cucumbers and melons. It is mainly used to control powdery mildew. It works by inhibiting the fungus's biosynthesis of important steroid molecules.

Description

Fenarimol was developed by Eli Lilly & Company around 1971.

Health Risk

Fenarimol can lead to increased growth of MCF7 breast cancer cells.

Synthesis

Fenarimol is made by the reaction of a diarylketon with an organolithium derived by halogen-metal exchange.

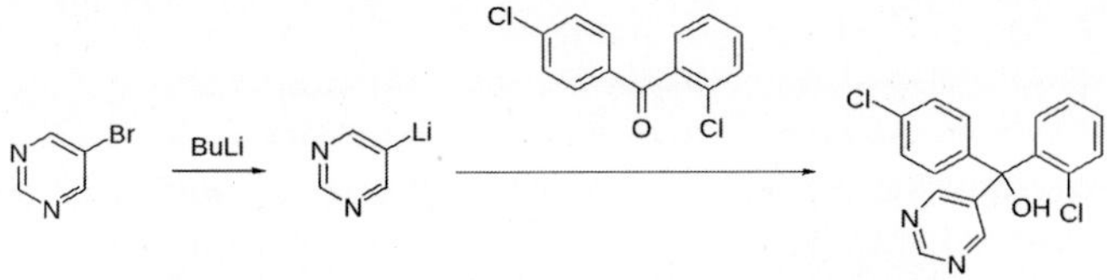

Fenpropimorph

Fenpropimorph is a morpholine-derived fungicide used in agriculture, primarily on cereal crops such as wheat.

Fentin Acetate

Fentin acetate is an organotin compound with the formula $(C_6H_5)_3SnO_2CCH_3$. It is a colourless solid that is used as a fungicide.

Flusilazole

Flusilazole (DPX-H6573) is an organosilicon fungicide invented by DuPont, which is used to control fungal infections on a variety of fruit and vegetable crops. It is moderately toxic to animals and has been shown to produce birth defects and embryotoxicity at high doses.

Fuberidazole

Fuberidazole (Chemical Formula: $C_{11}H_8N_2O$) is a chemical compound used in fungicides.

Hexachlorobenzene

Hexachlorobenzene, or perchlorobenzene, is a chlorocarbon with the molecular formula C_6Cl_6. It is a fungicide formerly used as a seed treatment, especially on wheat to control the fungal disease bunt. It has been banned globally under the Stockholm Convention on persistent organic pollutants.

Physical and Chemical Properties

HCB is a white crystalline solid that has negligible solubility in water (0.00000002 M) and variable solubility in different organic solvents. It is most soluble in halogenated solvents like chloroform (approx 0.03 M) less soluble in esters and hydrocarbons ((approx 0.020 M), and even less soluble in short chain alcohols (0.002-0.006 M). Its vapour pressure is 1.09×10^{-5} mmHg (1.45 mPa) at 20 °C. Its flash point is 242 °C and it sublimes at 322 °C.

Safety

Hexachlorobenzene is an animal carcinogen and is considered to be a probable human carcinogen. After its introduction as a fungicide in 1945, for crop seeds, this toxic chemical was found in all food types. Hexachlorobenzene was banned from use in the United States in 1966.

This material has been classified by the International Agency for Research on Cancer (IARC) as a Group 2B carcinogen (possibly carcinogenic to humans). Animal carcinogenicity data for hexachlorobenzene show increased incidences of liver, kidney (renal tubular tumours) and thyroid cancers.

Chronic oral exposure in humans has been shown to give rise to a liver disease (porphyria cutanea tarda), skin lesions with discolouration, ulceration, photosensitivity, thyroid effects, bone effects and loss of hair.

Neurological changes have been reported in rodents exposed to hexachlorobenzene. Hexachlorobenzene may cause embryolethality and teratogenic effects. Human and animal studies have demonstrated that hexachlorobenzene crosses the placenta to accumulate in foetal tissues and is transferred in breast milk.

HCB is very toxic to aquatic organisms. It may cause long term adverse effects in the aquatic environment. Therefore, release into waterways should be avoided. It is persistent in the environment. Ecological investigations have found that biomagnification up the food chain does occur. Hexachlorobenzene has a half life in the soil of between 3 and 6 years. Risk of bioaccumulation in an aquatic species is high.

Toxicology

- *Oral LD50 (rat):* 10,000 mg/kg
- *Oral LD50 (mice):* 4,000 mg/kg
- *Inhalation LC50 (rat):* 3600 mg/m^3

Material has relatively low acute toxicity but is toxic because of its persistent and cumulative nature in body tissues in rich lipid content.

Unique Exposure Incident

In Anatolia, Turkey between 1955 and 1959, during a period when bread wheat was unavailable, 500 people were fatally poisoned and more than 4,000 people fell ill by eating bread made with HCB-treated seed that was intended for agriculture use. Most of the sick were affected with a liver condition called porphyria cutanea tarda, which disturbs the metabolism of hemoglobin and results in skin lesions. Almost all breastfeeding children under the age of two, whose mothers had eaten tainted bread, died from a condition called "pembe yara" or "pink sore," most likely from high doses of HCB in the breast milk. In one mother's breast milk the HCB level was found to be 20 parts per million in lipid, approximately 2,000 times the average levels of contamination found in breast-milk samples around the world. Follow-up studies 20 to 30 years after the poisoning found average HCB levels in breast milk were still more than seven times the average for unexposed women in that part of the world (56 specimens of human milk obtained from mothers with porphyria, average value was 0.51 ppm in HCB-exposed patients compared to 0.07 ppm in unexposed controls), and 150 times the level allowed in cow's milk.

In the same follow-up study of 252 patients (162 males and 90 females, avg. current age of 35.7 years), 20–30 years postexposure, many subjects had dermatologic, neurologic, and orthopedic symptoms and signs. The observed clinical findings include scarring of the face and hands (83.7%), hyperpigmentation (65%), hypertrichosis (44.8%), pinched faces (40.1%), painless arthritis (70.2%), small hands (66.6%), sensory shading (60.6%), myotonia (37.9%), cogwheeling (41.9%), enlarged thyroid (34.9%), and enlarged liver (4.8%). Urine and stool porphyrin levels were determined in all patients, and 17 have at least one of the porphyrins elevated. Offspring of mothers with three decades of HCB-induced porphyria appear normal.

Hexachlorocyclohexa-2,5-dien-1-one

Hexachlorocyclohexa-2,5-dien-1-one, sometimes informally called hexachlorophenol (HCP) is an organochlorine compound which can be prepared from phenol. Despite the informal name, the compound is not a phenol but is a ketone; the informal name is derived from its method of preparation which includes phenol as a reagent.

Preparation

HCP is normally produced by electrophilic chlorination of phenol by chlorine gas in the presence of metal chloride catalyst, such as ferric chloride. It can also be produced by alkaline hydrolysis of chlorobenzenes at high temperature and pressure, by conversion of diazonium salts of chlorinated anilines, or by chlorination of phenolsulphonic acids and benzenesulphonic acids followed by removal of the sulphonic acid group.

Iprodione

Iprodione is an imidazole fungicide.

Application

Iprodion is used on crops affected by Botrytis bunch rot, Brown rot, Sclerotinia and other fungal diseases in plants. It is currently applied in a variety of crops: fruit, vegetables, ornamental trees and scrubs and on lawns. It is a contact fungicide that inhibits the germination of fungal spores and it blocks the growth of the fungal mycelium. It has been marketed under the brand name "Rovral" and "Chipco green" (both brands of Bayer CropScience). This chemical was developed originally by Rhône-Poulenc Agrochimie (later Aventis CropScience and in 2002 acquired by Bayer). Patents on iprodione are currently expired.

In 2010 Iprodione was launched in the United States as a nematicide with the brand name "Enclosure" by the agbiotech company DevGen.

Lime Sulfur

In horticulture, lime sulfur (British spelling lime sulphur) is a mixture of calcium polysulfides formed by reacting calcium hydroxide with sulfur, used in pest control. It can be prepared by boiling calcium hydroxide and sulfur together with a small amount of surfactant. It is normally used as an aqueous solution, which is reddish-yellow in colour and has a distinctive offensive odour.

A (New York State Agricultural Experiment Station) recipe for the concentrate is 80 lb. sulfur, 36 lb. high grade quicklime, and 50 gallons of water. 2.25:1 is the ratio (by weight) for compounding sulfur and quicklime; this makes the highest proportion of pentasulfide. If hydrated lime is used, an increase by 1/3 or more (to 48 lb. or more) might be used with the 80 lb. of sulfur. If the quicklime is 85%, 90%, or 95% pure, use 42 lb, 40 lb, or 38 lb; if impure hydrated lime is used, similarly increase its quantity. Avoid using lime that is less than 90% pure. Boil for an hour, stirring and adding small amounts of hot water to compensate for evaporation.

Use

Lime sulfur is sold as a spray for deciduous trees to control fungi, bacteria and insects living or dormant on the surface of the bark. Lime sulfur burns leaves so it is not as useful for evergreen plants.

Bonsai enthusiasts use undiluted lime sulfur to bleach, sterilise, and preserve deadwood on bonsai trees while giving an aged look. Rather than spraying the entire tree, as with the pesticidal usage, lime sulfur is painted directly onto the exposed deadwood, and is often coloured with a small amount of dark paint to make it look more natural. Without paint pigments, the lime-sulfur solution bleaches wood to a bone-white colour that takes time to weather and become natural-looking. Because the lime sulfur does not contact the leaves or needles, this technique can be used on evergreen trees as well as other types of trees. Diluted solutions of lime sulfur (between 1:16 and 1:32) are also used as a dip for pets to help control ringworm (fungus), mange and other dermatoses and parasites. (Undiluted lime sulfur is corrosive to the skin and can cause serious injury).

Safety

Lime sulfur reacts with strong acids (including stomach acid) to produce highly toxic hydrogen sulfide (rotten egg gas) and indeed usually has a distinct "rotten egg" odour to it. Lime sulfur is not extremely flammable but combustion produces highly irritating sulfur dioxide gas. Safety goggles and gloves should be worn while handling lime sulfur. Lime sulfur solutions are strongly alkaline (typical commercial concentrates have a pH over 11.5), and so it is corrosive to living things and can cause blindness if splashed in the eyes.

History

Lime sulfur is believed to be the earliest synthetic chemical used as a pesticide, being used in the 1840s in France to control grape vine powdery mildew *Uncinula necator*, which had been introduced from the USA in 1845 and reduced wine production by 80%. In 1886 it was first used in California to control San Jose scale. Commencing around 1904, commercial suppliers began to manufacturer lime sulfur; prior to that time, gardeners were expected to manufacture their own. By the 1920s essentially all commercial orchards in western countries were protected by regular spraying with lime sulfur. However by the 1940s, lime sulfur began to be replaced by synthetic organic fungicides which risked less damage to the crop's foliage.

Maneb

Maneb is a foliate fungicide and is a polymeric complex of manganese with the ethylene bis (dithiocarbamate) anionic ligand. It was included in a biocide ban proposed by the Swedish Chemicals Agency and approved by the European Parliament on January 13, 2009. It can be also used to create a toxin-based animal model of Parkinson's disease, usually in primates.

Metham Sodium

Metham sodium is a soil fumigant used as a pesticide, herbicide, and fungicide. It is one of the most widely used pesticides in the United States, with approximately 60 million pounds used in 2001. Metham sodium is the sodium salt of methyldithio-carbamate. Metham sodium can be prepared from methylamine, carbon disulfide, and sodium hydroxide; or from methyl isothiocyanate and sodium thiolate.

Upon exposure to the environment, metham sodium decomposes to form methyl isothiocyanate.

Methylisothiazolone

Methylisothiazolones are chemical compounds used as highly effective biocidal agents against fungal and bacterial species. They are widely used as a preservative in cosmetic industry, and as a pesticide, e.g. in paints and water tanks.

There are known bacterial species which exhibit increased resistance to the compound.

Most common compounds are:

- MCI (Methylchloroisothiazolinone, 5-Chlor-2-methyl-2,3-dihydroisothiazol-3-on, CAS 26172-55-4)
- MI (methylisothiazolone, 2-Methyl-2,3-dihydroisothiazol-3-on, CAS 2682-20-4).

Toxicity

Methylisothiazolone is highly corrosive and causes burns if use undiluted. Common usage rates in this preservative are .01%. The general public would not come into contact with undiluted MIT. Causes irritation of skin by contact if undiluted, and it is highly destructive of mucous membranes if applied undiluted.

It is lethal in low doses if absorbed via oral ingestion, inhalation or dermal contact. Known allergen in concentrations as low as 20 ppm, with prevalence in population between 3% and 8%. Suggested usage rate is .01% for lotions and usage rates. Maximum permissible dose for use in cosmetics was set to 15 ppm in year 1997. It is approved for use in over 25 countries and has a long history of safety data. Teratogenic risk: class D in Germany.

Monosodium Methyl Arsenate

Monosodium methyl arsenate (MSMA) is an arsenic-based herbicide and fungicide. It is an organic arsenate; but it is a less toxic organic form of arsenic, which has replaced the role of lead hydrogen arsenate in agriculture. It is one of the most common herbicides used on golf courses. It is typically used for control of grassy weeds such as crabgrass. Another common use is for burning the marked lines into grassy sports fields.

Some of its trade names include Ansar 170, Ansar 170L, Ansar 529, Ansar 529 HC, Ansar 6.6, Asazol, Bueno, Bueno 6, CCRIS 4676, Caswell No. 582, Daconate, Daconate 6, Dal-E-Rad, Dal-E-Rad 120, Drexar, Gepiron, HSDB 754, Herb-All, Herban M, Merge, Merge 823, Mesamate, Mesamate H.C., Mesamate-400, and Mesamate-600.

Myclobutanil

Myclobutanil is a triazole chemical used as a fungicide. It is a steroid demethylation inhibitor, specifically inhibiting ergosterol biosynthesis. Ergosterol is a critical component of fungal cell membranes.

Pentachloronitrobenzene

Pentachloronitrobenzene, typically abbreviated PCNB, is a registered fungicide formally derived from nitrobenzene. It is either an off-white or yellow solid, depending on its purity, with a musty odor.

Preparation

PCNB was originally synthesized in the laboratory in 1868. It was introduced to the agricultural world in the 1930s in Germany by Bayer AG as a substitute to mercurial pesticides. PCNB is prepared by chlorination of nitrobenzene at 60–70 °C in chlorosulfuric acid, with iodine as a catalyst. It can also be produced by the nitration of chlorinated benzenes. A side product of the synthesis of PCNB is hexachlorobenzene (HCB), which is considered as hazardous as PCNB.

$$5\ Cl_2 + C_2H_5NO_2 \rightarrow C_6Cl_5NO_2 + 5\ HCl$$

Main Reactions

Reaction with ethanol and potassium hydroxide yields pentachlorophenetole, indicating its high reactivity:

$$C_6Cl_5NO_2 + KOCH_2CH_3 \rightarrow C_6Cl_5OCH_2CH_3 + KNO_2$$

Although PCNB has a long shelflife, it is labile in soil, with a half life of 1.8 days.

It degrades to other metabolites, mainly reducing to pentachloroaniline (PCA), but also to pentachlorophenol (PCP) through hydrolysis and pentachlorothioanisole (PCTA).

Another metabolite is methyl pentachlorophenyl sulfide (MPCPS). Little information is available about the degradation mechanisms.

$$C_6Cl_5NO_2 + 6\ [H] \rightarrow C_6Cl_5NH_2 + 2\ H_2O$$

$$C_6Cl_5NO_2 + 6\ H_2O \rightarrow C_6Cl_5OH + HNO_2$$

Applications

PCNB is used as a fungicide to suppress the growth of fungi in various crops, such as cotton, rice, and seed grains. It is also used to prevent the formation of slime in industrial waters. In April 1993, PCNB was declared a hazardous air pollutant in the U.S. PCNB was reexamined for re-registration eligibility by the U.S. EPA in 2006 as part of the 1996 Food Protection Quality Act (FPQA) and as a result, use on a number of crops were ended or limited.

In August 2010, in response to the discovery of a potentially toxic metabolite in technical grade PCNB to be used in fungicide formulations, the sale of PCNB was halted by the U.S. EPA until the issue could be resolved.

PCNB is still used widely as a fungicide in other countries, such as China and Japan. Residual amounts of the compound and its metabolites are still found in crops today. The degradation products, PCA and PCTA have been found in farming soils and in river sediments.

Pentachlorophenol

Pentachlorophenol (PCP) is an organochlorine compound used as a pesticide and a disinfectant. First produced in the 1930s, it is marketed under many trade names. It can be found in two forms: PCP itself or as the sodium salt of PCP, which dissolves easily in water.

Synthesis

PCP can be produced by the chlorination of phenol in the presence of catalyst (anhydrous aluminium or ferric chloride) and a temperature of up to approximately 191 °C. This process does not result in complete chlorination and commercial PCP is only 84-90% pure.

The main contaminants include other polychlorinated phenols, polychlorinated dibenzo-p-dioxins, and polychlorinated dibenzofurans. Some of these species are even more toxic than the PCP itself.

Uses

PCP has been used as a herbicide, insecticide, fungicide, algaecide, and disinfectant and as an ingredient in antifouling paint. Some applications were in agricultural seeds (for nonfood uses), leather, masonry, wood preservation, cooling tower water, rope, and paper mills. Its use has declined due to its high toxicity and slow biodegradation.

There are two general methods for preserving wood. The pressure process method involves placing wood in a pressure-treating vessel where it is immersed in PCP and then subjected to applied pressure. In the non-pressure process method, PCP is applied by spraying, brushing, dipping, and soaking. Utility pole life, approximately 7 years for an untreated pole, is about 35 years for a treated pole.

Exposure

People may be exposed to PCP in occupational settings through the inhalation of contaminated workplace air and dermal contact with wood products treated with PCP. Also, general population exposure may occur through contact with contaminated environment media, particularly in the vicinity of wood treatment facilities and hazardous wastes sites.

In addition, some other important routes of exposure seem to be the inhalation of contaminated air, ingestion of contaminated ground water used as a source of drinking water, ingestion of contaminated food, and dermal contact with soils or products treated with the chemical.

Toxicity

Short-term exposure to large amounts of PCP can cause harmful effects on the liver, kidneys, blood, lungs, nervous system, immune system, and gastrointestinal tract. Elevated temperature, profuse sweating, uncoordinated movement, muscle twitching, and coma are additional side effects.

Contact with PCP (particularly in the form of vapor) can irritate the skin, eyes, and mouth. Long-term exposure to low levels such as those that occur in the workplace can cause damage to the liver, kidneys, blood, and nervous system. Finally exposure to PCP is also associated with carcinogenic, renal, and neurological effects. The U.S. Environmental Protection Agency Toxicity Class classifies PCP in group B2 (probable human carcinogen).

Monitoring of Human Exposure

Pentachlorophenol may be quantitated in plasma or urine as an index of excessive exposure. This is usually performed by gas chromatography with electron-capture or mass spectrometric detection. Since urine contains predominantly conjugated PCP in chronic exposure situations, prior hydrolysis of specimens is recommended. The current ACGIH biological exposure limits for occupational exposure to PCP are 5 mg/L in an end-of-shift plasma specimen and 2 mg/g creatinine in an end-of-shift urine specimen.

Absorption in Humans and Animals

PCP is quickly absorbed through the gastrointestinal tract following ingestion. Accumulation is not common, but if it does occur, the major sites are the liver, kidneys, plasma protein, spleen and fat. Unless kidney and liver functions are impaired, PCP is quickly eliminated from tissues and blood, and is excreted, mainly unchanged or in conjugated form, via the urine. Single doses of PCP have half-lives in blood of 30 to 50 hours in humans. Biomagnification of PCP in the food chain is not thought to be significant due to the fairly rapid metabolism of the compound by exposed organisms.

Releases to the Environment

PCP has been detected in surface waters and sediments, rainwater, drinking water, aquatic organisms, soil, and food, as well as in human milk, adipose tissue, and urine. As PCP is generally used for its properties as a biocidal agent, there is considerable concern about adverse ecosystem effects in areas of PCP contamination.

Releases to the environment are decreasing as a result of declining consumption and changing use methods. However, PCP is still released to surface waters from the atmosphere by wet deposition, from soil by run off and leaching, and from manufacturing and processing facilities. PCP is released directly into the atmosphere via volatilization from treated wood products and during production. Finally, releases to the soil can be by leaching from treated wood products, atmospheric deposition in precipitation (such as rain and snow), spills at industrial facilities and at hazardous waste sites.

After PCP is released into the atmosphere it decomposes through photolysis. The main biodegradative pathway for PCP is reductive dehalogenation. In this process, the compound PCP is broken down to tetrachlorophenols, trichlorophenols, and dichlorophenols. Another pathway is methylation to pentachloroanisole (a more lipid-soluble compound). These two methods eventually lead to ring cleavage and complete degradation.

In shallow waters, PCP is also quickly removed by photolysis. In deep or turbid water processes such as sorption and biodegradation take place. In reductive soil and sediments, PCP can be degraded within 14 days to 5 years, depending on the anaerobic soil bacteria that are present. However, adsorption of PCP in soils is pH dependent because it increases under acidic conditions and it decreases in neutral and basic conditions.

Pentachlorophenol by Country

New Zealand: PCP was used in New Zealand as a timber preservative and antisapstain treatment and it was no longer used after 1988.

It was also sold as a moss killer to the general public (by Shell, at least) in the form of a 115g/L aqueous solution and labelled as a poison.

United States

Since the early 1980s, the purchase and use of PCP in the U.S has not been available to the general public. Nowadays most of the PCP used in the U.S is restricted to the treatment of utility poles and railroad ties. In the United States, any drinking water supply with a PCP concentration exceeding the MCL, 1 ppb, must be notified by the water supplier to the public. Disposal of PCP and PCP contaminated substances are regulated under RCRA as F-listed (F021) or D-listed (D037) hazardous wastes.

2-Phenylphenol

2-Phenylphenol, or *o*-phenylphenol, is an organic compound that consists of two linked benzene rings and a phenolic hydroxyl group. It is a white or buff-coloured, flaky crystalline solid with a melting point of about 57 °C. It is a biocide used as a preservative under the trade names Dowicide, Torsite, Preventol, Nipacide and many others.

History

2-Phenylphenol (ortho-phenylphenol, OPP), and sodium o-phenylphenate, SOPP, were first evaluated by the 1962 JECFA for their use for the post-harvest treatment of fruits and vegetables to protect against microbial damage during storage and distribution in commerce.

Uses

The primary use of 2-phenylphenol is as an agricultural fungicide. It is generally applied post-harvest. It is a fungicide used for waxing citrus fruits. It is no longer a permitted food additive in the European Union, but is still allowed as a post harvest treatment.

It is also used for disinfection of seed boxes. It is a general surface disinfectant, used in households, hospitals, nursing homes, farms, laundries, barber shops, and food processing plants. It can be used on fibers and other materials. It is used to disinfect hospital and veterinary equipment. Other uses are in rubber industry and as a laboratory reagent. It is also used in the manufacture of other fungicides, dye stuffs, resins and rubber chemicals. 2-Phenylphenol is found in low concentrations in some household products such as spray disinfectants and aerosol or spray underarm deodorants. Eye contact can cause severe irritation and burns with possible eye damage. For some individuals, 2-phenylphenol can also irritate the skin. It is one of the chemicals that the Hyperactive Children's Support Group recommends be eliminated from the diet of children. The sodium salt of orthophenyl phenol, sodium orthophenyl phenol, is a preservative, used to treat the surface of citrus fruits to prolong shelf life.

Propamocarb

Propamocarb is a systemic fungicide used for control of soil, root and leaf disease caused by oomycetes. It is used by watering or spraying. Propamocarb is absorbed and distributed through crop's tissue.

Use

Propamocarb has fungicidal activity only against oomycetes.

Safety

Propamocarb has low general toxicity, and almost no teratogenicity or neurotoxicity for mammals. It is not a carcinogen nor mutagen. Propamocarb is not susceptible to formation of resistant diseases. It is fully metabolized by plants and aquatic bacteria in a few weeks, so it is not a major ecological threat. It possesses the risk of skin sensitization. Oral LD_{50} is 2900 mg/kg for male rats and 2000 mg/kg for female rats. In one study conducted on tobacco, cucumber and spinach, using propamocarb synthesized out of carbon C^{14} radionuclide researchers stated that propamocarb is decomposed down to carbon dioxide and then incorporated into plant's natural compounds, e.g. like amino acids.

Propiconazole

Propiconazole is a triazole fungicide, also known as a DMI, or demethylation inhibiting fungicide due to its binding with and inhibiting the 14-alpha demethylase enzyme from demethylating a precursor to ergosterol. Without this demethylation step, the ergosterols are not incorporated into the growing fungal cell membranes, and cellular growth is stopped. Propiconazole is used agriculturally on turfgrasses grown for seed and aesthetic or athletic value, mushrooms, corn, wild rice, peanuts, almonds, sorghum, oats, pecans, apricots, peaches, nectarines, plums and prunes. Propiconazole is a mixture of four stereoisomers and was first developed in 1979 by Janssen Pharmaceutica.

Pterulone

Pterulone is a fungal metabolite. It was initially isolated from the mycelium and liquid cultures of wood-decay fungus in the genus *Pterula*. The compound inhibits eukaryotic respiration by targeting the mitochondrial NADH:ubiquinone oxidoreductase.

Pyrazophos

Pyrazophos is a systemic fungicide which is used in orchards, vineyards and vegetable crops against powdery mildew.

QoI

Q_o inhibitors (Q_oI), or Quinone outside inhibitors are a group of fungicides used in agriculture. They represent the most important development made in fungicides by the chemicals industry. Q_oI are chemical compounds which act at the Quinol *outer* binding site of the cytochrome bc_1 complex.

Q_oI's are the resulting fusion of three fungicides families, the well-known family of strobilurins and two new families, represented by fenamidone and famoxadone. Some strobilurins are azoxystrobin, kresoxim-methyl, picoxystrobin, pyraclostrobin and trifloxystrobin. These fungicides are used on a wide range of crops, such as cereals, vines, pome fruits, cucurbits, tomatoes and potatoes. For example, they are used as fungicides for cereals, against *Erysiphe graminis f.sp tritici* responsible for the powdery mildew in wheat or against *Septoria tritici*, responsible for septoria leaf spot in wheat.

They are also commonly used for vine culture, against *Plasmopara viticola*, responsible for downy mildew or in oïdium treatment. All these fungicides are in the same cross-resistance group (same mode of action) and must be managed carefully to avoid the appearance of fungicide resistance. Some fungicide resistance has been observed in most crops (such as in the case of wheat powdery mildew), so the application of Q_0I products should respect effective rates and intervals to provides time and space when the pathogen population is not influenced by the product selection pressure.

Strobilurin

Strobilurins are a group of chemical compounds used in agriculture as fungicides. They are part of the larger group of QoI inhibitors, which act to inhibit the respiratory chain at the level of Complex III. Some strobilurins are azoxystrobin, kresoxim-methyl, picoxystrobin, fluoxastrobin, oryzastrobin, dimoxystrobin, pyraclostrobin and trifloxystrobin. Strobilurins represented a major development in fungus-based fungicides. They were extracted from the fungus *Strobilurus tenacellus*. They have a suppressive effect on other fungi, reducing competition for nutrients; they inhibit electron transfer in mitochondria, disrupting metabolism and preventing growth of the target fungi.

Strobilurins are mostly contact fungicides with a long half time as they are absorbed into the cuticle and not transported any further.

Tebuconazole

Tebuconazole is a triazole fungicide used agriculturally to treat plant pathogenic fungi. Though the U.S. Food and Drug Administration considers this fungicide to be safe for humans, it may still pose a risk. It is listed as a possible carcinogen in the United States Environmental Protection Agency Office of Pesticide Programs carcinogen list with a rating of C (possible carcinogen). Its acute toxicity is moderate. According to the World Health Organization toxicity classification, it is listed as III, which means slightly hazardous.

Due to the potential for endocrine-disrupting effects, tebuconazole was assessed by the Swedish Chemicals Agency as being potentially removed from the market by EU regulation 1107/2009.

Tiabendazole

Tiabendazole (INN, other names include TBZ, thiabendazole and the trade names Mintezol, Tresaderm, and Arbotect) is a fungicide and parasiticide.

Uses

Fungicide: It is used primarily to control mold, blight, and other fungally caused diseases in fruits (e.g. oranges) and vegetables; it is also used as a prophylactic treatment for Dutch elm disease. Use in treatment of aspergillosis has been reported.

Parasiticide

As an antiparasitic, it is able to control roundworms (such as those causing strongyloidiasis), hookworms, and other helminth species which attack wild animals, livestock and humans.

Angiogenesis Inhibitor

Genes responsible for the maintenance of cell walls in yeast have been shown to be responsible for angiogenesis in vertebrates. Tiabendazole serves to block angiogenesis in both frog embryos and human cells. It has also been shown to serve as a vascular disrupting agent to reduce newly established blood vessels. Tiabendazole has been shown to effectively do this in certain cancer cells.

Pharmacodynamics

TBZ works by inhibition of the mitochondrial, helminth-specific enzyme, fumarate reductase, with possible interaction with endogenous quinone.

Other

Medicinally, thiabendazole is also a chelating agent, which means that it is used medicinally to bind metals in cases of metal poisoning, such as lead, mercury or antimony poisoning. In dogs and cats, thiabendazole is used to treat ear infections. Thiabendazole is also used as a food additive, a preservative with E number E233 (INS number 233). For example, it is applied to bananas to ensure freshness, and is a common ingredient in the waxes applied to the skins of citrus fruits. It is not approved as a food additive in the EU, Australia and New Zealand.

Safety

The substance appears to have a slight toxicity in higher doses, with effects such as liver and intestinal disorders at high exposure in test animals (just below LD_{50} level). Some reproductive disorders and decreasing weaning weight have been observed, also at high exposure. Effects on humans from use as a drug include nausea, vomiting, loss of appetite, diarrhea, dizziness, drowsiness, or headache; very rarely also ringing in the ears, vision changes, stomach pain, yellowing eyes and skin, dark urine, fever, fatigue, increased thirst and change in the amount of urine. Carcinogenic effects have been shown at higher doses.

Timorex Gold

Timorex Gold is a biofungicide based on a plant extract of Melaleuca alternifolia, which was developed by Biomor, a member of the Stockton Group. Timorex Gold is effective against a wide range of plant-pathogenic fungi, including Oomycetes, Ascomycetes, and Basidomycetes, on numerous crops, including vegetables, herbs, grapevines, fruit trees, and banana. Timorex Gold has demonstrated high efficacy against Black Sigatoka (Black leaf streak disease in banana). Plants damaged by Black Sigatoka have a significant reduction in the photosynthesizing area of the leaf, and fruit yied losses can reach 50% through premature maturation.

Black Sigatoka is caused by Mycosphaerella fijiensis (Morelet), and is considered the most damaging and costly disease of commercial banana and plantain. It originated in the Pacific region, was reported in Fiji in 1963, and since then has reached other banana producing countries. Control of this disease may require as many as 70 treatments per year using synthetic fungicides.

Timorex Gold's special formulation contains multiple components, mostly terpenes and their alcohols, which have been proven antiseptic, fungicide and bactericide properties. The unique and powerful potential of this plant extract is highly effective for control in agricultural crops while significantly reducing the chemical risk to operators and the environment.

History

In 2003, scientists at Biomor,s member of the Stockton Group Research and Development Innovation Centre, commenced research on the use of *Melaleuca alternifolia* in agriculture for the management and control of bacterial and fungal diseases. Studies dealing with non-plant pathogens confirmed that the fungicidal and anti-microbial activity of *Melaleuca alternifolia* resulted from its ability to disrupt the permeability barrier in membrane structures. In yeast cells and isolated mitochondria, á-pinene and â-pinene, plant extract of "Melaleuca alternifolia" destroy cellular integrity, inhibit respiration and ion transport processes and increase membrane permeability. Timorex Gold has proven to be very effective against a broad spectrum of diseases found in vegetables, herbs, grapevines, bananas, and fruit trees.

Registration

The active substance of Timorex Gold was approved by the European Union to be included in the positive list of the EU, in Annex 1 of Directive 91/414/EEC for the registration of pesticides.

Toxicology

The active ingredient of Timorex Gold is a natural product that is classified as a low-risk substance for which Maximum Residue Limits (MRL) are not required in Europe. Stockton is in the process of obtaining similar exemptions in other areas of the world such as North America and Asia.

Organic Accreditation

Timorex Gold is recognized to be a product that is used in both conventional and organic agriculture. Timorex Gold meets the requirements of European Union, NOP and JAS and is certified by the following entities:

- Israeli Bio-Organic, Agriculture Association (I.B.O.A.A.)
- International Federation of Organic Agriculture Movements (IFOAM))
- BCS Öko-Garantie
- State of Israel Ministry of Agriculture & Rural Development
- Control Union (CU) Certifications

Triamiphos

Triamiphos (chemical formula: C12H19N6OP) is a chemical compound used in fungicides.

2,4,6-Tribromophenol

2,4,6-Tribromophenol (TBP) is a brominated derivative of phenol. It is used as a fungicide, as a wood preservative, and an intermediate in the preparation of flame retardants.

Production

Although natural TBP has been identified in ocean sediments as a metabolite of marine fauna, the commercial product is prepared industrially. In 2001, the production volume of TBP was estimated to be 2500 tonnes/year in Japan and 9500 tonnes/year worldwide. TBP can be prepared by the controlled reaction of elemental bromine with phenol:

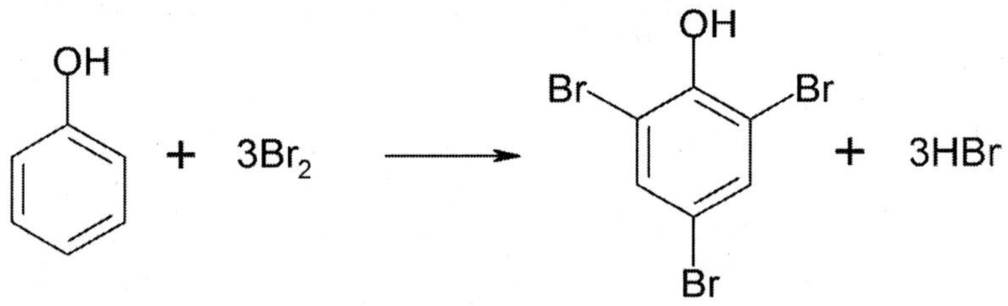

Uses

The predominant use of TBP is as an intermediate in the preparation of flame retardants such as brominated epoxy resins. TBP is reacted with sodium hydroxide to form the sodium salt, which is used as a fungicide and wood preservative.

The bismuth salt is the active ingredient in Xeroform dressing. Microbial metabolism in products treated with TBP is known to produce 2,4,6-tribromoanisole.

Tributyltin Oxide

Tributyltin oxide (TBTO), or bis(tri-*n*-butyltin)oxide, is an organotin compound chiefly used as a biocide (fungicide and molluscicide), especially a wood preservative. Its chemical formula is $C_{24}H_{54}OSn_2$. It has the form of a thin, colourless to pale yellow liquid with melting point -45 °C, boiling point 180 °C, and slight water solubility (20 ppm). It is combustible and soluble in organic solvents.

It is available under names AW 75-D, Bio-Met TBTO, Biomet, Biomet 75, BTO, Butinox, C-SN-9, Hexabutyl-distannoxane and others.

Tributyltin Oxide is a Potent Skin Irritant.

Tributyltin compounds had been used as marine anti-biofouling agents. Concerns over toxicity of these compounds (some reports describe biological effects to marine life at a concentration of 1 nanogram per litre) have led to a worldwide ban by the International Maritime Organization. It is now considered a severe marine pollutant and a Substance of very high concern by the EU.

Triclocarban

Triclocarban (TCC), or 3,4,4'-trichlorocarbanilide, is a substance with anti-bacterial and anti-fungal properties that is used in disinfectants, soaps and other household products.

The mechanism of action involves disruption of cellular membranes, and the chemical may also involve inhibition of the enzyme enoyl-acyl carrier protein reductase (ENR). This enzyme is absent in humans but is essential in building cell membranes of many bacteria and fungi.

Safety

TCC has low toxicity (LD_{50} >5000 mg/kg) and its absorption across human skin is very low.

One preliminary study found that high concentrations of TCC enhance the biological activity of the endogenous hormone testosterone *in vitro* and in a male rat model.

Validamycin

Validamycin is a antibiotic and fungicide produced by *Streptomyces hygroscopicus*. It is used as an inhibitor of trehalase. It is used for the control of sheath blight of rice and damping-off of cucumbers.

Vinclozolin

Vinclozolin (trade names: Ronilan, Curalan, Vorlan, Touche) is a common dicarboximide fungicide used to control diseases, such as blights, rots and molds in vineyards, and on fruits and vegetables such as raspberries, lettuce, kiwi, snap beans, and onions. It is also used on turf on golf courses. Two common fungi that vinclozolin is used to protect crops against are botrytis cinerea and sclerotinia scleotiorun. First registered in 1981, vinclozolin is widely used but its overall application has declined. As a pesticide, vinclozolin is regulated by the United States Environmental Protection Agency (U.S. EPA). It has gone through a series of tests and regulations in order to evaluate the risks and hazards to the environment and animals. Among the research, a main finding is that vinclozolin has been shown to be an endocrine disruptor with antiandrogenic effects.

Use in the United States

Vinclozolin is manufactured by the chemical company BASF and has been registered for use in the United States since 1981. The following is a compilation of data indicating the national use of vinclozolin per crop (lbs AI/yr) in 1987: apricots, 124; cherries, 3,301; green beans, 13,437; lettuce, 24,779; nectarines, 1,449; onions, 829; peaches, 15,203; plums, 163; raspberries, 3,247; and strawberries, 41,006. In 1997, two applications totaling 285 pounds each, were applied to kiwifruit in California to prevent the gray mold and soft rot caused by Botrytis cinerea. In general, the United States has seen an overall decline in the national use of vinclozolin. In 1992, a total of approximately 135,000 pounds were used. However, in 1997 this number dropped to 122,000 and in 2002 it was down to 55,000 pounds.

Preparation and Application

The following chemical reactions are used to make vinclozolin:

- Methyl vinyl ketone + sodium cyanide + 3,5-dichloroaniline + phosgene (cyanohydrin formation/nitrile hydrolysis/amide formation/phosgenation).
- Vinclozolin is also prepared by the reaction of 3,5-dichlorophenyl isocyanate with an alkyl ester of 2-hydroxy-2-vinylpropionic acid; ring closure is achieved at elevated temperature.

Vinclozolin is then formulated into a dry flowable or extruded granular. It can be applied by through the air (aerial), through irrigation systems (chemigation), or by ground equipment. Vinclozolin is also applied to some plants, such as decorative flowers, as a dip treatment where the plant is dipped into the fungicide solution and then dried. It is also common to spray a vinclozolin solution using thermal foggers in greenhouses.

History of Regulations

All pesticides sold or distributed in the United States must be registered by EPA. To qualify for registration, scientific studies need to show that the pesticide can be used without putting people or the environment at risk. The law also requires that pesticides which were first registered before November 1, 1984, be reregistered so that they can be retested using the now more advanced scientific methods. Because vinclozolin was released in 1981, it has gone through a preliminary registration, and also a subsequent reregistration. Below is a list of the history of regulations for vinclozolin:

- A Data Call-In (DCI), requiring data on product and residue chemistry, toxicity, environmental fate, and ecological effects, have been requested in 1991, 1995 and 1996.
- Agricultural Data Call-In (AGDCI), to estimate what happens to workers after they apply the fungicide, was issued in 1995.
- In April 1997, BASF proposed a new use for vinclozolin and thus all risks were reevaluated under the Food Quality Protection Act (FQPA).
- In June 1998 the EPA put a new safety factor on vinclozolin. BASF voluntarily cancelled vinclozolin use on some fruits and turf.
- On July 18, 2000 the EPA established that certain meat, bean and dairy products have a 3 year time-limited tolerances for vinclozolin and its metabolites. This led to a phase-out of most domestic food uses for vinclozolin. In September 2000, the EPA received objections to the issued tolerance for beans.
- Other measures that BASF has taken to reduce risks: cancel use on decorative plants and set up new restrictions on turf which state that the application to turf restricted to golf courses and industrial sites.
- BASF has proposed to immediately eliminate or phase-out uses such that only use on canola and turf will remain after 2004.

Safety

There are certain precautions that workers are advised to use while handling vinclozolin. Mixers and loaders must wear clothing that completely covers the skin. They must also wear a chemical resistant apron and gloves, along with protective eyewear. Those who use the air blast equipment must be in a closed cab that meets the definition in the Worker Protection Standard for Agricultural Pesticides for dermal protection.

Means of Exposure

The EPA has examined whether people can be exposed to vinclozolin or its metabolites by means of dietary (food and water), non-dietary and occupational exposure. In general, fungicides have been shown to circulate through the water and air, and it possible for them to end up on untreated foods after application. Consumers alone cannot easily reduce their exposure because fungicides are not removed from produce that is washed with tap water.

It has been determined that people may be exposed to residues of vinclozolin and its metabolites containing the 3,5-dichloroaniline moiety (3,5-DCA) through diet, and thus tolerance limits have been established for each crop. Although vinclozolin is not registered for use by homeowners, it is still possible for people to come in to contact with the fungicide and its residues. For example, golfers playing on treated golf courses, and families playing on sod which has been previously treated may be at risk for exposure. Occupationally, workers can be exposed to vinclozolin while doing activities such as loading and mixing.

Chapter 4

Environmental and Health Impacts

Antiandrogen

As part of the reregistration process, the EPA reviewed all toxicity studies on vinclozolin. The main effect induced by vinclozolin is related to its antiandrogenic activity and its ability to act as a competitive antagonist at the androgen receptor. Vinclozolin can mimic male hormones, like testosterone, and bind to androgen receptors, while not necessarily activating those receptors properly.

There is evidence that vinclozolin itself binds weakly to the androgen receptor but that at least two of its metabolites are responsible for much of the antiandrogenic activity. When male rats were given low dose levels (>3 mg/kg/day) of vinclozolin, effects such as decreased prostate weight, weight reduction in sex organs, nipple development, and decreased ano-genital distance were noted.

At higher dose levels, male sex organ weight decreased further, and sex organ malformations were seen, such as reduced penis size, the appearance of vaginal pouches and hypospadias. In the rat model, it has been shown that the antiandorgenic effects of vinclozolin are most prominent during the developmental stages. In utero, this sensitive period of fetal development occurs between gestation days 16-17. Embryonic exposure to vinclozolin can influence sexual differentiation, gondal formation, and reproductive functions. In bird models, vinclozolin and its metabolites were shown in vitro and in vivo to inhibit androgen receptor binding and gene expression. Vinclozolin caused reduced egg laying, reduced fertility rate, and a reduction in successful hatches. Androgens also play a role in puberty, and it has been shown an antiandrogen like vinclozolin can delay pubertal maturation. Antiandrogenic toxins are also known to alter sexual differentiation and reproduction in the rabbit model.

Male rabbits exposed to vinclozolin in utero or during infancy did not show a sexual interest in females or did not ejaculate. Since the androgen receptor is widely conserved across species lines, antiandrogenic effects would be expected in humans.

Progesterone and Estrogen Effects

Besides androgens, vinclozolin has been shown to affect other steroid hormone receptors, such as those of progesterone and estrogen. Just as with androgens, the timing of the exposure to vinclozolin determines the magnitude of the effects related to these hormones.

In a study with rats, in vitro research showed the ability of two vinclozolin metabolites to bind to the progesterone receptor. However, the same study in vivo using adult male rats showed no effects. When mice experienced vinclozolin exposure in utero, male offspring exhibited up-regulated estrogen receptor and up-regulated progesterone receptor. In females, vinclozolin down-regulated expression of estrogen receptors and up-regulated progesterone receptor expression. This result causes virilization and the feminization of males and masculinization of females.

Transgenerational Effects

Vinclozolin has been demonstrated to have trangenerational effects, meaning that not only is the initial animal affected, but effects are also seen in subsequent generations. One study demonstrated that vinclozolin impaired male fertility not only in the first generation that was exposed in utero, but in males born for three generations and beyond.

Furthermore, when affected males with were mated with normal females, some of the offspring were sterile and some had reduced fertility. After three generations, male offspring continued to show low sperm count, prostate disease and high rates of testicular cell apoptosis., Other studies conducted experiments where rat embryos were exposed to vinclozolin during sex determination. F1 (first generation) vinclozolin treated males were bred with F1 vinclozolin treated females. This pattern continued for three generations. The initial F0 mother was the only subject that was directly exposed to doses of vinclozolin. F1-F4 generation males all showed an increase in the prevalence of tumors, prostate disease, kidney disease, test abnormalities and immune failures when compared to the control group. F1-F4 females also showed an increased incidence of tumors and kidney disease. Furthermore, transgenerationally transmitted changes in mate preference and anxiety behaviour have also been observed in rats following exposure to vinclozolin. It has been reported that these transgenerational reports correlate with epigenetic changes, specifically, an alteration in DNA methylation in the male germ line.

Links to Cancer

The EPA has classified vinclozolin as a possible human carcinogen. Vinclozolin induces an increase in leydig cell tumors in rats. The 3,5-DCA metabolite is thought to possess a mode of tumor induction based on its similarity to p-choroaniline.

Environment

In general, fungicides can enter surface water in turf runoff, potentially contaminating drinking water sources. Laboratory test indicate that vinclozolin is easily broken down and dissipates in the environment with the help of microbes, soil metabolism, and transport with water. Vinclozolin has several metabolites that are the products of its degradation: Metabolite B, 3,5-DCA and metabolite E. Only 3,5 DCA appears to resist further degradation. In terrestrial field dissipation studies conducted in various states, vinclozolin dissipated with a half-life between 34 to 94 days. Half-lives including residues can reach up to 1,000 days. Residues may accumulate and be available for future crop uptake.

Future Implications

Since the phase-out of vinclozolin, farmers are faced with fewer options to control gray and white mold. The New York State Agricultural Experiment Station has carried out efficacy trials for gray and white mold. Research has showed potential alternatives to vinclozolin. Rifloxystrobin (Flint), iprodione

(Rovral), and cyprodinil plus fludioxonil (Switch) control gray mold. Thiophanate-methyl (Topsin M) was as effective as vinclozolin in controlling white molds. Switch was the most promising alternative to vinclozolin for controlling both gray and white mold on pods and for increasing marketable yield.

Xanthene

Xanthene (9*H*-xanthene, 10*H*-9-oxaanthracene) is a yellow organic heterocyclic compound. Its chemical formula is $C_{13}H_{10}O$. It is soluble in diethyl ether. Its melting point is 101-102 °C and its boiling point is 310-312 °C.

Xanthene is used as a fungicide and it is also a useful intermediate in organic synthesis. Derivatives of xanthene are commonly referred to collectively as xanthenes, and among other uses are the basis of a class of dyes which includes fluorescein, eosins, and rhodamines.

Xanthene dyes tend to be fluorescent, yellow to pink to bluish red, brilliant dyes. Many xanthene dyes can be prepared by condensation of derivates of phthalic anhydride with derivates of resorcinol or 3-aminophenol.

Zinc Borate

Zinc borate is an inorganic compound, a borate of zinc. It is a white crystalline or amorphous powder insoluble in water. Its toxicity is low. Its melting point is 980 °C.

Variants

Several variants of zinc borate exist, differing by the zinc/boron ratio and the water content:

- Zinc borate Firebrake ZB ($2ZnO \cdot 3\ B_2O_3 \cdot 3.5H_2O$), CAS number 138265-88-0
- Zinc borate Firebrake 500 ($2ZnO \cdot 3\ B_2O_3$), CAS number 138265-88-0
- Zinc borate Firebrake 415 ($4ZnO \cdot B_2O_3 \cdot H_2O$), CAS number 149749-62-2
- ZB-467 ($4ZnO \cdot 6B_2O_3 \cdot 7H_2O$), CAS number 1332-07-6
- ZB-223 ($2ZnO \cdot 2B_2O_3 \cdot 3H_2O$), CAS number 1332-07-6

The hydrated variants lose water between 290–415 °C.

Uses

Zinc borate is primarily used as a flame retardant in plastics and cellulose fibers, paper, rubbers and textiles. It is also used in paints, adhesives, and pigments. As a flame retardant, it can replace antimony(III) oxide as a synergist in both halogen-based and halogen-free systems. It is an anti-dripping and char-promoting agent, and suppresses the afterglow. In electrical insulator plastics it suppresses arcing and tracking. In halogen-containing systems, zinc borate is used together with antimony trioxide and alumina trihydrate. It catalyzes formation of char and creates a protective layer of glass. Zinc catalyzes the release of halogens by forming zinc halides and zinc oxyhalides. In halogen-free system, zinc borate can be used together with alumina trihydrate, magnesium hydroxide, red phosphorus, or ammonium polyphosphate. During burning the plastics, a porous borate ceramics is formed that protects the underlying layers. In presence of silica, borosilicate glass can be formed at plastic burning temperatures.

Zinc borate is used in polyvinyl chloride, polyolefins, polyamides, epoxy resins, polyesters, thermoplastic elastomers, rubbers, etc. It is also used in some intumescent systems.

Zinc borate has synergistic effect with zinc phosphate or barium borate as a corrosion inhibitor pigment.

Zinc borate acts as a broad-spectrum fungicide in plastics and wood products.

Zinc borate can be used as a flux in some ceramics. In electrical insulators it improves the ceramics properties. Nanopowder zinc borate can be used for the applications above, and also for improving the frictional properties of lubricating oils.

Zineb

Zineb is a foliate fungicide and is a polymeric complex of zinc with the ethylene bis(dithiocarbamate) anionic ligand. Its uses include control of downy mildews, rusts and redfire disease. It is produced by reacting ethylene bis(dithiocarbamate) sodium salt, nabam, with zinc sulfate. This can be carried out by mixing nabam and zinc sulfate in the spray tank. In the US it was once registered as a "General Use Pesticide", however all registrations were voluntarily cancelled following an EPA special review. It continues to be used in many other countries.

Ethylene bis(dithiocarbamate) with Manganese instead of Zinc, called Maneb, was included in a biocide ban proposed by the Swedish Chemicals Agency and approved by the European Parliament on January 13, 2009.

Ziram

Ziram or Zinc Dimethyldithiocarbamateis a carbamate fungicide. It is used in agriculture either as a spray applied to the leaves to control fungal infections. It may also be used for seed treatment to improve the chances of germination. The chemical is also used in the manufacture of rubber.

Fungicide Use in the United States

Almonds

Alternaria Leaf Spot: Symptoms of Alternaria leaf spot appear as lesions with tan spots on the leaves. The centres of these lesions become black with fungal sporulation. This infection can lead to tree death within 3–4 years of the first serious outbreak. Orchards in high humidity areas result in the largest yield loss, often in excess of 50%. Yield loss tends to rise every year as the tree becomes weaker each year after infection. Three fungicide applications can achieve 60–80% control of leaf spot.

Anthracnose

Anthracnose was not seen on California almonds until the early 1990s. By 1996 it was widespread and causing severe yield losses throughout the state. Typical losses in 1996 were 10–15% of the almond crop with severely affected crops incurring losses of 25%. Under wet conditions, orange spore masses are produced and appear as visible droplets. Lesions on mature fruit are rusty orange and gum profusely. Once the diseased fruit die they become mummies that remain on the tree. The pathogen overwinters in these mummies. 80–90% control can be achieved by applying fungicides to protect the crop before rains begin. The California Department of Pesticide Regulation has estimated that without the fungicides to control anthracnose the state's almond production would drop 15–30%.

Brown Rot

Damage from brown rot occurs several years after the infection strikes. The primary symptom is fruiting spur loss.

Brown rot was first discovered on California almonds in the late 19th century and currently affects most almond-producing areas of California. Brown rot can be controlled using fungicides through bloom in order to protect the flower parts from brown rot attacks. Experiments have demonstrated that 44% of twigs were infected with brown rot when crops were left untreated compared to only 4% when the crop was treated.

Crown and Root Rot

Crown and root rot of almonds is caused by at least 14 different *Phytophthora* species. The risk of root or crown infection is greatest during cool to moderate temperatures with prolonged or frequent soil saturation. A tree infected with *Phytophthora* can either undergo a period of slow decline that may last years or it can suddenly collapse and die in spring with the advent of warm weather. Eventually, leaves drop, terminal shoots die back, and death of the tree follows. Once in the root or crown the infection may extend into the crown, trunk, or branches. Currently, crown and root rot are a problem affecting 20% of California's almond orchards with potential yield losses of 50%.

Green Fruit Rot

Green fruit rot can be found throughout virtually all almond-producing regions of California. Green rot is typically controlled by fungicides applied to control other fungal diseases that occur during blooming. It is only when cooler temperatures and heavy moisture is present that almond growers are recommended to make fungicide applications specifically for the disease. When left untreated, green fruit rot can cause up to a 10% yield loss.

Leaf Blight

Leaf blight of almonds was first discovered in 1950, and by 1983 it had spread throughout the Sacramento Valley. Almond leaf blight is characterized by the death of leaves throughout the summer. The fungus interferes with water conduction in the leaf. In fall and winter, dormant buds are killed by an extension of the lesion and in the spring, flowers are killed. Repeated attacks of leaf blight, though rarely killing more than 20% of the leaves, ultimately reduce tree vigor. Experiments have shown that the use of ziram and captan can reduce the incidence of leaf blight by 75–80% in treated trees.

Rust

Almond rust is a disease of almond trees that is characterized by angular yellow leaf spots on upper leaf surfaces and rusty brown masses of spores on lower leaf surfaces. It was first discovered in the upper Sacramento Valley area of California but now has subsequently spread throughout both the Sacramento and San Joaquin valley orchards. It typically emerges in the summer and fall seasons and has been shown to cause rapid and extensive defoliation of trees. Treatments of maneb and sulfur in the early season months (typically spring and summer) have been shown to significantly reduce incidence and severity of rust outbreaks.

Scab

Almond scab was first documented in almond orchards in the 1950s. However it did not become a major disease until sprinkler irrigation became popular in the 1980s. By the early 1990s almond scab could be found throughout California almond orchards. Scab infects leaves, fruit, and twigs in almonds causing dark spots to form. Scab lesions look greasy and oily. The major concern with almond scab is partial or complete defoliation of the tree. As with leaf blight use of captan and ziram has been shown to reduce total incidence of almond scab infection.

Shot Hole

Shot hole affects almonds by disrupting both the leaves and the fruit of the plant. It can result in premature nut ripening and plant defoliation. In some occasions, especially when rain persists in the springtime, complete defoliation and tree weakening can occur. Estimates have shown that approximately 80% of California's almond acres are infected with this fungus. A four year research project by the Almond Board of California determined that production losses from shot hole could range from 50 to 75%. Captan in one to three applications has been used as the primary fungicide to combat this disease. It has been shown to reduce lesions per fruit by 96%.

Apples

Spraying for control of fungal diseases of apples started in the U.S. some time between 1880 and 1905.

Apple Scab

Apple scab is caused by the fungus *Venturia inaequalis*. Mating among different strains of the fungus occurs shortly after leaf fall with spores developing in the fallen leaves during the winter. Spring rains cause spores to be forcibly discharged; they can be carried long distances by air currents to flowers, leaves, or young fruit. The spores then continue to develop and are released over a period of 5–9 weeks. These spores germinate and penetrate the outer layers of the plant, causing infection. The fungus grows beneath the cuticle and eventually ruptures to form dark green lesions. The number of lesions per leaf can be as few as 1–2 or there can be hundreds.

The USDA's National Agricultural Pesticide Impact Assessment Program estimated that 100% of eastern apple orchards are stricken with apple scab and that without fungicide treatments yield losses would be as high as 90%.

In the western orchards 52% are infected and apple yield loss can be as high as 22%. However with the use of fungicides experiments have shown heavy reductions in the percentage of infected apples. One study was able to reduce the incidence from 77% to 2%.

Bitter Rot

Bitter rot is a major disease in the Southeast U.S. in the summer months when the weather is hot and damp. The organism has a short incubation period and as a result epidemics of bitter rot can develop rapidly. The presence of the disease is first indicated by the very small light brown sunken spots beneath the apple skin. As the fungus grows and invades more of the apple tissue, the area becomes engorged until the entire apple is rotted. Around the start of the 20th century Bordeaux mixture was the primary technique for controlling the disease; in the 1940s growers shifted to synthetic chemical use. Without these fungicides is it estimated that apple yield loss due to bitter rot would be as high at 90%.

Black Pox

This disease primarily affects the southeastern portion of the United States and is known to infect the cultivars Rome Beauty, Grimes Golden, Delicious, York Imperial, and Golden

Delicious. Lesions on twigs are well defined, conical, shiny black swellings and on the fruit itself they are black, spherical, slightly sunken spots. Severely affected leaves may die within 2 to 3 weeks of infection. Infected branches will grow poorly, lose their leaves early, and die. Black pox can be controlled with the same fungicide sprays that are used to treat scab.

Blossom-End Rot

The first symptom of blossom-end rot is soft, wet, and reddish discolourations that appear in the late summer months. As the rot stops growing it will begin to dry out and appear sunken. The affected fruit will often drop prematurely. Fungicide experiments have shown reduced incidence, from 5% to less than 0.5%, with treatment.

Brooks Fruit Spot

Brooks fruit spot is a minor disease of the apple and is found mostly throughout the Northeast and Mid-Atlantic regions. As the fruit first emerges the disease appears as dark green lesions on the fruit of the apple. As the apple begins to emerge these blemishes grow and change to purple or green. Brooks fruit spot is usually controlled with fungicides applied during the early cover-spray period. One study showed that apple orchards that were not treated with fungicides had 87% of their fruit infected with brooks fruit spot while only 1–6% of trees that were treated with fungicides showed symptoms of the disease.

Fire Blight

Fire blight was first described in New York in the late 18th century, and moved west with settlers, becoming established throughout North American apple production areas by the early 20th century. While fire blight has always been a concern in eastern apple production, severe outbreaks in the west in recent years have caused growers there to adopt more consistent and vigorous monitoring and management programs as well. The disease can affect every part of the tree, from the fruit to the trunk. Infected trees may die within months or can linger for years with severely reduced yields. In addition to the scorched appearance of the plant parts that gives the disease its name, plant tissues infected with the bacteria will exude milky or reddish-brown ooze. Initially copper sprays were used for fire blight control in the 1930s but this method had limited success. In the 1950s streptomycin and oxytetracycline showed high success in controlling fire blight in comparison to copper. Since then streptomycin sprayed two to three times during blooming phase has become the treatment of choice.

Powdery Mildew

Powdery mildew is a common fungal infection of apples and can occur in almost any apple growing climate. The spores from fungi preserved over the winter are released from the unfolding leaves of the buds. The spores, carried by the wind, infect leaves, blossoms, and fruit. The fungus spreads until it covers the whole leaf and then grows down twigs, covering them with a gray felt. This results in aborted blossoms, reduced finish quality, and reduced yield. The USDA has estimated that 40% of apple orchards in the East and 50% of apple orchards in the West are infected with powdery mildew. Without control of the fungus yields would drop by 65%.

Quince Rust

Quince rust infects the fruit of the apple trees but does not affect the leaves. Quince rust spores infect cedar trees and create cylindrical galls from which emerge spore horns the following spring.

These galls may produce spores for up to twenty years. Quince rust is economically important primarily when an extended wetting period with a mean temperature above 10°C occurs between the tight cluster and late pink bud stages. Under these conditions, economic losses may occur throughout large geographic areas. Experiments with fungicide sprays have been shown to provide complete control of rust.

White Rot

White rot gets its name from the soft, watery, and light-coloured rotted fruit that is left over after an apple is infected. The fungus survives from season to season in the dead bark and mummified fruit of the apple tree. Spores can survive in the dead bark for up to six years. Growers are advised to treat for white rot once the sugar content of the fruit reaches approximately 10%. USDA estimates that 20% of apple orchards in eastern states are infected with the fungus and that without fungicide use yield losses would be 65%.

Artichokes

California produces 100% of all artichoke crops in the U.S. 84% of this acreage is located near the cool, moist, coastal area of Monterey Bay. The climate there is ideal for artichoke production.

Asparagus

Crown Rot: Crown rot in asparagus was first reported in California in 1938 when Californian growers began to report a slimy, orange-yellow asparagus tissue with soft lesions causing post harvest losses of 20–30%. As lesions grow they eventually collapse and shrivel. In the 1980s it was shown that fungicidal treatments of metalaxyl increased yields by 80%. Today mefenoxam is the primary fungicide used to treat crown rot.

Purple Spot

Purple spot was first observed in the United States in the 1980s. Yield loss due to this disease is delayed because the damage to fern growth does not affect current yields but instead affects the future crop. Damage results in the defoliation of needles, reducing the flow of carbohydrates to the roots and subsequently lowering the next year's yield by up to 52%. In the spring spores produced from the previous year's infected crops are spread by water and wind to new host plants.

The primary control of purple spot was EBDC fungicides until processors began to reject the use of EBDCs. As a result since 1990 exemptions for chlorothalonil and tebuconazole in Michigan have been granted by the EPA for use on asparagus. The fungicides are applied at the end of the harvest and chlorothalonil has been shown in experiments to reduce purple spot by 99% and increase yields by 36%. This increase in yield as a result of purple spot control has been shown to translate into a net return of $200–400 per acre for asparagus growers.

Rust

Asparagus rust was first reported in the U.S. in 1896. It began on the asparagus crops of mostly northeastern states and began to move westward across the country- For each successive year the disease was discovered in new western areas until in 1902 it was reported in California for the first time. Research to control the disease began with Bordeaux mixture experiments around the start of the 20th century. Copper and sulfur dust were tried as well but their success was very limited. Sulfur was the only one to have much success at all. Rust resistant strains of asparagus were cultivated but by the 1940s and 1950s these strains of asparagus were no longer providing adequate levels of resistance. In the 1950s zineb and mancozeb were developed as the primary means to control rust. Zineb reduced infection by 85% and mancozeb by 97%. By 1989 EBDC fungicides had become the most common fungicidal means to control rust on asparagus. However after the EPA almost cancelled the use of EBDCs on asparagus many processors decided they would not accept EBDC-treated asparagus despite the EPA allowing its use after review. This meant that many growers were left with no registered fungicide to use on asparagus. Since 1990 the EPA has granted exemptions for tebuconazole, myclobutanil, and triadimefon for use on asparagus.

Bananas

Hawaii is the primary state in the U.S. where bananas are grown. The average yield is 15,000 pounds/acre. Approximately 600-800 plants are grown per acre. Planting occurs year round. Banana bunches are ready for harvest 12 to 15 months after initial planting. On a banana plantation, plants can be seen at all stages of vegetative growth and fruit maturity year-round. Bananas can be harvested any day of the year.

Black Sigatoka

Like yellow sigatoka, black sigatoka, was first documented in the Sigatoka valley of Fiji. It was first recorded in 1964 and being more virulent tended to displace yellow sigatoka in banana crops. Therefore yellow sigatoka is rarely found in locations where black sigatoka occurs. Black sigatoka infection appears on the leaves of crops during the unfurling. Sigatoka spores will incubate on the leaves for up to six days before penetrating the leaf. After this the infection will continue to colonize for a week before the plant exhibits symptoms. The initial symptoms are small spots on the undersides of the leaves. These appear 10–15 days after infection and grow until they appear as black streaks on the leaves. This is what gives black sigatoka its alternate name of black leaf streak. These streaks can dry out and collapse in less than a day. This affects growth and yield of bananas by reducing the total photosynthetic area of the leaf. However the largest effect on yields is through the toxins produced by black sigatoka that causes a premature ripening of the bananas. These prematurely ripened fruit cannot be sold and must be discarded.

Under normal conditions in Hawaii it is estimated that without control of black sigatoka yield losses would be 30%. Sigatoka is controlled with a combination of protectant and systemic fungicides in order to prevent resistance from developing. Mancozeb is applied as the primary protectant and fenbuconazole, tebuconazole, and azoxystrobin are applied as systemic fungicides. These fungicides are applied during summer.

Yellow Sigatoka

Yellow sigatoka derives its name from the Sigatoka valley in Fiji. This was where the disease was first documented in 1912. Over the course of the first half of the 20th century the disease spread to all major banana-producing countries of the world. It was first seen in Trinidad in 1934 and quickly spread to the mainland and up through Central America. The disease can spread spores over 1,000 miles with wind currents. By 1936 experiments with Bordeaux mixture were developed to control the disease. The disease was first discovered in Hawaii in 1958.

Barley

The major barley producing states of the United States are Idaho, Minnesota, Montana, and North Dakota. These states make up 70% of barley production in America. This barley is principally used in malting and brewing. It is also in some areas used as a feed grain.

Barley Scab

Outbreaks of barley scab (also known as fusarium head blight) usually occur when there are high moisture conditions during the flowering stages of barley development. The infected crops become bleached in colour, shrunken, and discoloured and will grow pink spore masses on the barley heads. The disease has the potential to destroy a crop within weeks. The shrunken kernels can become so light that the wind can blow them away from the rest of the crop. In addition, barley scab produces toxins that severely reduce the value of the harvested barley. This toxin, deoxynivalenol, also known as vomitoxin because of its tendency to induce vomiting at high levels, will pass through the brewing process into the beer. When the beer is opened the toxin will cause the liquid to overflow from the bottle or can. Because this is not commercially desirable, barley producers receive severe price discounts when vomitoxin is present.

Currently control of barley scab is seen as a high priority, resulting in the U.S. National Wheat and Barley Scab Initiative. In 2005 alone North Dakota State experts estimated that scab cost the North Dakota farm economy $162 million and in 1993 it cost North Dakota, South Dakota, and Minnesota $1 billion. Fungicide trials in 2004 were promising. Results demonstrated that fungicide use reduced scab by 77–85%, reduced vomitoxin levels by 49–69%, and increased yields by 10–14%.

Stripe Rust

Stripe rust is a recent fungus to arrive in North and South America. The disease was first observed in Colombia in 1975 and is believed to have been brought over from Europe. In the U.S. it was first discovered in Texas in 1991 and by the mid-90s it could be found throughout barley crops in the western states. Stripe rust is now established in California, Idaho, Oregon, and Washington and all other barley areas of the Pacific Northwest are considered highly susceptible. Between 1996 and 1998 in Oregon yield losses of 25–50% were reported and in California between 15 and 30% were recorded. The state of Idaho estimates that without fungicide use yield losses would be 40%, as opposed to 5% with fungicide application.

Blueberries

Blueberries are susceptible to a number of fungal diseases. These have been shown by the USDA to reduce yields by 25–60% if left uncontrolled.

Alternaria Leaf Spot

Leaf spot was first identified in North Carolina in the early 1970s when it was found to be causing widespread fruit decay in blueberries. The disease is especially harmful to growers because symptoms of the disease do not develop until May, about a month before the June harvest. When the fruit ripens it becomes covered in a greenish fungal growth and can become leaky. Because the fungus develops late in the harvest, it can be effectively controlled by applying fungicides at early bloom and then every two weeks until harvest.

Botrytis Blight

Botrytis blight is a grey mold fungus that was first observed on New Jersey blueberries in 1924 and by the 1950s was considered the most dangerous disease of the Northwest. USDA estimates that 95% of Northwestern blueberry crop and 40% of eastern blueberry fields are infected with botrytis blight. Yield losses within these fields with uncontrolled blight are estimated to be 30–40%. In most cases infections in the blossoms of the plant results in ovary mortality which prevents development of the blueberry itself.

Fusicoccum Canker

Fusicoccum canker is a disease of the blueberry that is mostly limited to Michigan blueberry producers. It was first discovered in the 1960s and has been an annual problem of the lower portion of the peninsula. When left uncontrolled the disease has been shown to reduce yields by 30%. Fusicoccum canker is primarily controlled using captan. Research has shown that regular applications can reduce infectious cankers by 82–95%.

Mummyberry

This fungal infection strikes the blueberry plant in the early spring. Over the winter the mummyberry infection incubates in the mummified blueberry fruit that has fallen on the ground and in the spring cup-shaped structures of the fungus begin to grow. These structures will then eject spores for nine days at an average of sixty thousand spores per day. These spores are carried by the wind to the twigs and flowers of developing blueberry plants. The fungus then infects the young leaves and flowers. This is infection stage 1. The secondary infection stage occurs when the infected leaves and flower grow conidia, these conidia release spores which will actually "fertilize" healthy blueberry flowers causing the plant to grow a thick fungus sclerotium that helps the fungus to survive the winter. These berries then fall to the ground where they serve as the hosts for the next year's mummberry infection. If the infection were left uncontrolled the USDA estimates that blueberry yields would fall an average of 20 to 50%, without control, the state of Maine estimated their yields would drop 25% and Michigan has reported that they would lose 25–57% of the crop.

Experiments with fungicide applications have reduced the incidence of mummyberry at harvest from 21–24% to 0.4% and increased yield by up to 34%.

Phytophthora Root Rot

Phytophthora root rot was first reported in the blueberry plant in 1963. In plants surveyed in North Carolina it was found that 40% of plants were infected. Symptoms of root rot include yellowing or reddening of leaves, heavy defoliation, and stunted plant growth. Today current USDA estimates are that 80% of eastern blueberry acres are infected with root rot and it has the potential to reduce yields by up to 25%.

Pseudomonas Blight

Pseudomonas blight is a bacterial disease that invades the blueberry plant early in the season when tissue is damaged by freezing temperatures. The bacteria invade the tissue of the plant and spread throughout the plant as long as it remains cool and damp. As the infection spreads thin strands of ooze appear on the surface of the tissue. When temperatures rise the disease will retreat until fall when the colder weather allows it to return. Copper is primarily used to control the bacteria population.

Ripe Rot

Ripe rot, the more common name for anthracnose, is a disease localized mostly to the northern and southern blueberry-growing regions. However is has been spreading east and it is now estimated to infect 30% of the eastern blueberry growing regions. Much like leaf spot, ripe rot symptoms appear as the blueberry begins to ripen and mature. Once the blueberry has ripened, it softens and orange spores will be exuded. This disease can be especially costly because if a diseased berry showing no symptoms is put in a container of untainted berries it can very quickly infect the entire container. Control of this disease is achieved with fungicidal applications every 1–2 weeks.

Septoria Leaf Spot

Septoria leaf spot infects the blueberry plant through spores ejected from infected leaves and stem lesions left on the ground from the previous harvest season. These spores infect the crop and small white/tan leaf lesions form on the plant and the stem in early May. These lesions continue to develop and multiply through September. Severe infection causes defoliation of the plant. This will lower fruit yield as the nutrient supply is reduced. In weaker infections where defoliation does not occur the loss of photosynthetic area will reduce yields as well.

Use of fenbuconazole has been shown to be successful in controlling leaf spot. If applied at harvest it will help prevent leaf spot from forming the next year. It has been shown to average 45% greater yields in the next harvest.

Cabbage

Alternaria Leaf Spot: Alternaria leaf spot affects the cabbage plant by causing severe spotting and discolouration to the head of the plant.

It is only a superficial effect to the plant but because of the black colour of the leaf spots the marketability of a severely infected cabbage crop is low. Without control leaf spot is estimated to be able to reduce yields by up to 50%. Alternaria leaf spot is controlled using the same treatment as downy mildew: applications of azoxystrobin, copper, chlorothalonil, mancozeb, and maneb beginning midseason.

Black rot

Due to its ability to spread rapidly and destroy entire fields of crops black rot can be considered the most serious disease of the cabbage plant. Under the right conditions one plant has the ability to spread the disease to an entire field and cause 100% yield loss. The disease thrives in areas with heavy rainfall and warmer temperatures. The bacteria invade the plant through the water the plant absorbs and once inside spreads rapidly. It obtains its name because it will blacken and clog the veins of the plant making it impossible to keep nutrients moving through the plant. Copper is the primary treatment for black rot control.

Clubroot

Clubroot gets its name from its primary symptom: enlargement of the plant roots. The abnormally large roots are unable to properly absorb water and nutrients from the ground and the result is cabbages that are undersized, wilted, and discoloured.

If left uncontrolled yield losses due to clubroot can be as high as 50%. The only fungicide shown to be effective in treating clubroot is PCNB.

Downy Mildew

Downy mildew left uncontrolled has the ability to reduce yields by up to 55% and infest up to 80% of plants.[64] Downy mildew is primarily controlled through the use of several fungicides applied during midseason. The fungicides used include azoxystrobin, copper, chlorothalonil, mancozeb, and maneb. Farmers in Florida who regularly use fungicides have been shown to reduce yield losses due to downy mildew to as low as 2%.

Carrots

Bacterial Blight: Uncontrolled bacterial blight has been shown to cause yield losses up to 20%. Symptoms include small, dry, and brittle yellowish-brown spots on the plant and stalks covered in bacterial ooze. The primary treatment of bacterial blight is applications of copper before the crop is fully mature.

Cavity Spot

Cavity spot differs from the other diseases of the carrot in that symptoms are only visible on the roots of the plant and therefore it is impossible to tell if a plant is infected from above ground. The primary symptom of cavity spot is small half-inch lesions on the roots of the plant. This disease is treated with metalaxyl which allows for almost complete control of the disease. Without metalaxyl studies have found 50–60% of the carrot crop will be infected; after metalaxyl application, incidence is under 1%.

Powdery Mildew

Powdery mildew is one of the more recent diseases of carrots. It was first discovered in 1975 in the United States in the warmer climates of California and Texas.

Powdery mildew will cover the carrot with a film of white fungal growth which does not kill the plant, but stunts the growth of the carrot. These weaker plants are nearly impossible to harvest mechanically.

Collards

There are three major disease of collards: Alternaria, downy mildew, and black rot. Alternaria symptoms are small brown lesions on the surface of the collard leaf. Downy mildew produces small yellow fungal spots on the leaf that cause the plant to wither and die.

Finally, black rot gets its name because its primary symptom is the fungus will infect the veins of the plant and turn them black. These black veins eventually collapse and the plant dies. All three of these diseases are controlled with copper sprays.

Cotton

Pythium and Rhizoctonia: Pythium and Rhizoctonia are both seedling diseases of the cotton plant. They can be devastating to cotton plants because they develop when seedlings are unable to overcome the damage being caused by the invading fungus. The diseases often kill seeds before they grow. However if the seeds do manage to sprout they demonstrate symptoms almost immediately.

The stalks of the cotton plant will be weak, brown sunken lesions will appear on the stem, and the plant will often die before producing or will have very poor yields.

Growers who plant early are especially susceptible because cooler temperatures and damper weather are conducive to these diseases. Therefore growers who plant early are advised to use in-furrow fungicides for control: PNCB and iprodione on Rhizoctonia and etridiazole and mefnoxam on pythium. Recently, azoxystrobin has proven effective in controlling both rhizoctonia and pythium.

Rust

Rust primarily attacks plants in Arizona, New Mexico, and Texas. The disease attacks during the summer months and causes yellow or orange spots to grow on the leaves of the cotton plant. These spots can cause defoliation, premature openings, broken stems, and reduced yield. Crops with severe outbreaks have been known to incur losses of up to 50%. This disease is controlled with applications of mancozeb.

Garlic

Rust: Early symptoms of garlic rust are small yellow spots on the leaves that soon expand until the leaf tissue shatters and visible pustules emerge.

Diseased bulbs lose their protective dry outer skin, preventing photosynthesis and leaving the garlic prone to shattering when mechanically harvested. In trials tebconazole and azoxystrobin have been shown to provide 50% higher yields in treated crops.

Grapes

Black Rot: Black rot is a disease estimated to affect 95% of all eastern grape vineyards. The USDA estimates it could reduce yields by 85% if left untreated. Black rot is spread through spores left in the mummied grapes infected the year before. Upon infection the grapes shrivel up and turn black. These grapes do not fall from the vine but stay attached, and over winter spores grow within the infected grape.

When summer comes the rain re-moistens the dried grape and causes it to swell. Spores discharge from the swollen grape into the air and infect the healthy crop. Prior to fungicide development for black rot, in conditions favourable to the disease losses of 70 to 100 percent of the crop were common. Today however applications of fungicides have been able to give almost complete control of the disease. In many vineyards infections can be reduced from 95% down to 1%.

Botrytis Bunch Rot

Cold and wet climates favour this disease, which is why it is often found in the coastal vineyards of California and New York. In the eastern vineyards this disease is estimated to infect 30% of grapes while in California it can infect up to half.

This can reduce yields by 40–60%. Much like black rot, bunch rot overwinters in the berries and infects the next season's crop through spore ejections in the summer. Originally Bordeaux mixture was used to controlled botrytis bunch rot but the use of synthetic chemicals is now the preferred method.

Phomopsis Cane and Leaf Spot

Cane and leaf spot is a disease found on large areas of the East coast but rarely in California. The USDA estimates it affects 75% of the acres of eastern vineyards, with the potential to reduce yields up to 15%.

The disease can result in lesions, defoliation, and berry rot. However the same fungicide sprays used to control black rot done early in the season have been shown to control cane and leaf spot.

Powdery Mildew

Powdery mildew was first reported in European grape vineyards in the mid 1850s. It could reduce yields by up to 80%. In France alone it reduced wine production from a billion gallons to only two hundred million gallons. It was as a result of this disease's rapid spread throughout the continent that the first fungicide, sulfur, was discovered.

The success of sulfur was so great that the grape industry was sparked in California when it was seen that disease could now be controlled and grapes were easy to grow. Uncontrolled powdery mildew rots the grapes from the inside out. Powdery mildew slows the growth of the grape skin so that the inside pulp grows at a faster rate and bursts through. These split grapes dry up and rot. The grapes that do not split open are of much lower quality than normal grapes. Their acidity is much higher and as a result the wines taste bitter and off to most wine drinkers. In order to control powdery mildew the same treatment that was recommended in the late 19th century is still used today: regular applications of sulfur every 1–2 weeks.

Green Beans

Alternaria: Alternaria was first reported on green beans in Florida in 1951 after a hundred acres were lost to the disease in one season. By the 1970s it had spread up the coast to New York. The disease causes lesions on the skin of the green bean, killing the tissue and giving the plant a moldy appearance. Trials have shown that applications of chlorothalonil can reduce alternaria by up to 85%.

Bacterial Blight

Bacterial blight is endemic to green bean crops grown east of the Rocky Mountains. During ideal conditions (usually humid weather) the disease can inflict losses as high as 60%. Symptoms include lesions resembling burn marks that gradually grow larger. In some states incidence of the disease can be as high as 100%. Trials have shown that properly timed copper sprays can reduce incidence by 90%.

Root Rot

Root rot is a common disease of Florida green beans and can reduce crop yields there by up to 75% if untreated. It was not until the 1950s that an effective fungicide was found to control root rot. Trials in Florida in the 1950s showed that applications of PCNB were able to reduce incidence to 3%. PCNB is still the primary treatment today.

Powdery Mildew

Powdery mildew appears on the green bean plant as a powdery white mold on the surface of the green bean leaf. This can grow and cover the entire plant.

Yield losses occur because the damage makes the crop not commercially viable. Powdery mildew can be controlled with applications of sulfur. The original trials in the 1930s showed that incidence could be reduced from 87% to 2%.

Rust

Bean rust is a disease of the green bean that causes rust-coloured pustules on the leaves. These pustules eventually rupture and eject thousands of spores into the air to be carried to other plants. Each individual spore can produce a fresh pustule on a plant within a week; therefore untreated rust can rapidly reach epidemic levels.

When leaves become infected with rust they rapidly die causing defoliation. Research in the 1930s showed that sulfur was an effective treatment for rust and since then yield losses in treated fields have been reduced by 60%.

White Mold

Before effective fungicides were developed white mold caused significant hardship for green bean growers. White mold is considered a resilient disease. After infecting the plant white mold produces black structures called sclerotia that fall to the soil and can survive for over five years until the conditions are right for infection again. Since white mold affects plants very late in the growing season farmers would often harvest early to avoid losing much of their crop. However this usually resulted in a 30% loss of crop. Research in the 1970s eventually discovered that fungicides applied every 20 days reduced white mold incidence to 0% in treated areas.

Hazelnuts

Eastern Filbert Blight: Eastern filbert blight is caused by the fungus Anisogramma anomala and is indigenous to the Northeast United States. The fungus causes a small canker on the native American Hazelnut, Corylus americana. However, on the introduced and commercially important European Hazelnut, Corylus avellana, it causes a lethal disease.

The cankers caused by EFB slowly expand and kill the tree over the course of several years if diseased tree limbs are not removed in time.

Research with chlorothalonil and tebuconazole in the 1990s showed that 4–5 applications of the fungicides on a two-week schedule when vegetative buds are breaking dormancy is ideal for control of EFB. The state of Oregon has estimated that if left uncontrolled EFB could eliminate half of the state's hazelnuts within ten years.

Hot Peppers

New Mexico accounts for roughly 65% of all U.S. hot pepper production.

Bacterial Spot

Bacterial spot is spread from plant to plant through water, wind, and plant contact. Once infected, the leaves of the plant are targeted by the disease.

The disease causes severe spotting of the pepper and kills the leaves. This is a twofold problem because the defoliation results in the pepper being discoloured by sunscald. Research for bacterial spot treatment has shown that copper sprays have been able to increase marketable yields by 50% in treated fields.

Powdery Mildew

Powdery mildew causes a white powdery growth to cover the leaves of the hot pepper plant. These leaves may drop prematurely and as with bacterial spot can then cause sunscald. This disease has been known to cause yield losses as high as 50–60% in untreated fields. The prime fungicide used to control powdery mildew is myclobutanil.

Lettuce

Bottom Rot: Bottom rot symptoms consist largely of deep reddish-coloured lesions on the leaves that slowly spread from leaf to leaf until the entire head is rotted. Trials with vinclozolin have demonstrated effectiveness in controlling bottom rot. Research indicated that when treated, crop yields rise about 36%.

Downy Mildew

Downy mildew is an extremely common disease of the lettuce plant. This is largely due to the disease's quick germination period—three hours—and the lettuce plant being susceptible at all growth stages. One study estimated that in untreated Florida fields 75% of lettuce was infected with downy mildew and considered unsalable. Research with maneb has shown the ability to reduce the number of lesions per 10 plants from 187 to about one. EBDC and fosetyl-al have also been shown effective in Florida trials that reduced unmarketable heads to 2%.

Lettuce Drop

Lettuce drop is another resilient disease of the lettuce plant. It can be found in all lettuce-growing regions of the U.S. and is able to survive in the soil for up to a decade after infecting a field.

The fungus appears on the plant as a white mass on the leaf that slowly turns into a hard black growth. These growths release spores that infect the lettuce plant with pathogens that slowly move up the stalk, killing tissue as they travel. Use of

DCNA, iprodione, and vinclozolin about a month before harvest has been shown to significantly reduce lettuce drop in treated fields. In treated California fields fungicides are able to reduce yield losses as low as 5%.

Powdery Mildew

Powdery mildew attacks lettuce plants by robbing them of the nutrients within the leaf. The fungus appears as a white growth on the upper surface of the leaf where it uses small tubes to pull nutrients out of the plant. This causes severe discolouration and eventual defoliation of the plant. Applications of sulfur are recommended as the primary way to control powdery mildew on lettuce.

Onions

Downy Mildew: Downy mildew is an extremely fast-developing disease in the onion. Studies have observed up to 40% of a field being infected in less than three days. The first symptoms of the disease are white and purple growths on the surface of the plant leaves.

Eventually these leaves die. This in turn destroys the plant's photosynthetic abilities, stunting growth and sometimes killing the plant. Untreated downy mildew can reduce yields by 65%. Downy mildew is controlled with the same foliar fungicides as leaf blight: azoxystrobin, copper, mancozeb, maneb, and mefenoxam.

Botrytis Leaf Blight

Leaf blight occurs in areas with hot and humid weather. The spores land on onion leaves, penetrate the skin, and rapidly kill the leaf. This causes a severe reduction in bulb size, as high as 50%, and can happen in less than a week. Applications of foliar fungicides can be used to control leaf blight. These include azoxystrobin, chlorotholonil copper, mancozeb, maneb, and mefenoxam.

Purple Blotch

Symptoms of purple blotch are primarily small white lesions that grow larger as they turn purple and brown. From these lesions spores are produced that cause additional lesions and also spread to other plants. However to develop, the disease requires very wet conditions for an extended period of time. Fungicide-treated fields have been found to have yields roughly 20–35% higher than untreated fields.

Seed Rot

Yield loss due to seed rot can be as high as 30% in heavily infected acres. The disease infects the onion seed and will cause the onions roots to be gray and mushy. These roots and plants rapidly decay and decompose, killing the plant. Mefenoxam is recommended as treatment for seed rot.

Smut

Smut is a tough disease of the onion because it can last for years in the soil as spores. When the onion crop is planted the spores germinate and enter the fresh seeds.

When the plant grows lesions form on the leaf and any bulbs that form will be covered in black spots. Most of the time infected seeds will die within 3–5 weeks of planting. Mancozeb is recommended for treatment of smut.

Peanuts

Cylindrocladium Black Rot: This disease infects any part of the plant below ground, so it often targets the roots; pegs or pods that are below ground are also vulnerable. Infected parts turn black and die.

Often the infection spreads and kills the entire root system, effectively killing the plant. Originally the disease was localized to North Carolina and Virginia, but recently Florida and Georgia have begun experiencing higher incidences of the rot. Recent research has shown tebuconazole to significantly suppress the disease and increase yields by up to 1,500 pounds per acre.

Leaf Spot, Early

Spots first appear on the upper surface of lower leaves as faint brown to black pinpoint dots. As the dots enlarge to become brown to dark-brown circular spots, a yellow "halo" generally develops as a border around each spot.

This disease is frequently seen as early as June 1 in both North Carolina and Virginia. Spots with an irregular shape can also develop on leaf petioles and plant stems. Defoliation and reduced yield at harvest can result if this disease is not controlled by fungicide sprays. Leaves that fall to the soil surface may trigger epidemics of certain soilborne diseases such as Southern stem rot.

Leaf Spot, Late

Like early leafspot, this disease can also cause defoliation, reduced yield, and increased incidence of certain soilborne diseases such as Southern stem rot. As the name implies, late leafspot is most prevalent during the later part of the growing season. Spots are generally darker in colour than early leafspot and commonly show no yellow halo.

Rust

Peanut rust is unique because it is not a native disease to the U.S. and has not been able to survive beyond a single season when introduced. Airborne spores are annually introduced from other peanut-producing nations. Rust causes reduced peanut pod fill and can potentially defoliate the entire crop if the conditions are right. Originally copper and sulfur were used to control rust, but they were only partially effective. It was not until the development of chlorothalonil and tebuconazole that an effective treatment for peanut rust was found.

Sclerotinia Blight

Scletotinia blight was first discovered in Virginia in the early 1970s but soon spread to the peanut producing states of North Carolina, Oklahoma and Texas. By 1982 it was considered to be one of the more significant diseases of the peanut. In the late 1990s it was estimated that 70% of Virginia peanut land was infected with sclertotinia blight. The disease attacks the plant by producing a white mold on the stems that slowly consumes the stem, causing pods to detach early and leaving them undersized. Pod loss can be as high as 50% in infected areas. In Virginia a five year experiment saw yield increase of 35% in fields treated with fluazinam.

Stem Rot

Stem rot, also known as white mold or southern blight, can be found in just about any peanut producing area in the U.S. The most notable symptom of this disease is the white cottony fungal growth that envelops the entire plant. This mold will appear midseason when the foliage has covered the row middles.

This damages the pods and creates a rotted texture. In the 1950s PCNB was found to be somewhat effective at preventing white mold; however it was only able to reduce incidence by 15%. It was not until the 1990s when tebuconazole was introduced that a fungicide was able to provide effective disease control. Tebuconazole is able to reduce incidence of white mold by 80–90%.

Web Blotch

Web blotch is currently found in all major peanut growing states in the south. It can be a highly damaging disease. In ideal conditions web blotch can cause yield loss as high as 50%. The first sign that a plant is infected with web blotch is small tan blotches on plant leaves. The leaf will become brittle and the plant will defoliate. Chlorothalonil and tebuconazole are effective treatments for reducing web blotch incidence.

Pecans

Pecan Scab: Pecan scab is the deadliest disease of the pecan. In the right conditions—high humidity and dampness—up to 100% crop reduction can occur if left uncontrolled. As a result fungicides are used on 85–90% of all pecan acreage in southeastern states annually. Fenbuconazole, propiconazole, and TPTH are the primary fungicides used in the southeast for scab control. In the southwest however fungicides are rarely used because the dry growing conditions are not conducive to growth of pecan scab.

Pistachios

Panicle and Shoot Blight: In California, since its discovery, panicle and shoot blight have become major diseases of the pistachio crop. The warm, wet weather in the California pistachio acreage has proven to be optimal for the disease. In 1998 alone total lost production was estimated to be around 20 million pounds. The only areas that have been found resilient to the disease are Kern County and parts of the San Joaquin valley. Symptoms include dark lesions and shoots from the buds that are stunted and black. Eventually the leaves on the shoots wither and die and the fruit clusters subsequently collapse. Currently azoxystrobin and chlorothalonil are used for blight control.

Raspberries

Botrytis Fruit Rot: This disease infects plants from berries left mummified from last year's crop. When the flowers open, spores from these mummified berries are expelled and infect the plant. This disease can be very costly because no symptoms are exhibited by the plant until harvest time, and then symptoms appear very rapidly. The infected berries quickly develop a layer of gray mold (gray mold is also an alternate name for botrytis fruit rot), begin to leak fluid, and release more spores to cause additional infections. Since the disease is such a late and rapidly occurring disease preventative fungicidal sprays must be used for control. Recent research has demonstrated an 83% reduction of gray mold in treated harvests.

Phythophthora Root Rot

If the conditions are right root rot can inflict yield losses up to 75% on a raspberry crop. Symptoms of the disease are destruction of root tissue and wilted primocanes and floricanes. Metalaxyl and fosetyl-al applied once in the fall or early spring have been shown to provide effective control of the disease.

Yellow Rust

The primary modes of dispersal of yellow rust are wind and the splashing of raindrops, making wet conditions ideal for its spread. This means the Pacific Northwest is very susceptible to yellow rust. The symptoms of yellow rust are typically yellow pustules of the leaves that cause the leaves to fall from the plant. The fruit will also often die before it ripens. In favourable conditions entire fields can easily incur 100% yield loss. Oregon has estimated that treating fields with fungicides can provide 98–100% control of yellow rust, whereas uncontrolled yellow rust could reduce yields up to 25%.

Rice

Sheath Blight: Sheath blight, caused by the fungus *Rhizoctonia solani*, is one of the deadlier diseases of rice. Up to 50% of southern United States rice could be infected with sheath blight each year. Unlike most infections it does not produce spores and instead must be spread from contact through plants, rainwater, or the soil. It is a resilient disease that can survive for years in the soil even when there is no rice crop present. Symptoms include lesions on the sheath of the plant, which feed off the nutrients within the plant. These lesions give rise to brown fungus balls which, if the plant dies from nutrient starvation, fall to the ground.

The following season, heavy rains may allow the balls to float to other plants, which they can then infect. Primary yield loss is therefore attributable to plant death and lower yields from the surviving plants due to the nutrient loss.

Yield losses can be as high as 42%. Research data has demonstrated that one application of azocystrobin at panicle differentiation can control sheath blight for almost the entire season.

Kernel Smut

Kernel smut can be found throughout soils in the Mid South where spores are able to survive up to two years without a crop being planted. When a crop is planted the disease infects the rice kernels during early development and replaces the starch inside with black kernel smut spores. These spores swell the plant until the grain eventually cannot contain the spores and bursts, releasing spores into the air. These spores either infect the current crop or land on the soil to infect the next year's crop.

The disease harms the plant because when rice infected with kernel smut is harvested it often breaks during the milling process and is rendered useless.

Research in Arkansas has recently shown that yield losses is about 10% when rice is left untreated. Propiconazole has been found to provide the best fungicidal control, offering 85–95% reduction in kernel smut.

False Smut

False smut damages the rice plant by infecting the ovaries of the rice kernel in its early development. Once inside the disease takes over the ovary and replaces it with spores that burst, producing a large orange ball between the glumes. These galls not only are covered in spores that spread the disease but when harvested result in rice that must be cleaned before it can be sold. Use of propiconazole can reduce false smut by as much as 75%.

Soybeans

Anthracnose: Anthracnose tends to develop late in the harvesting season. The disease infects soybean by attaching to the surface of the plant and penetrating the tissue wall. From here anthracnose is able to spread spores throughout the plant, killing branches, pods, and leaves. Once the plant is infected the disease causes defoliation and reduced quality and quantity of soybean seeds. Foliar fungicides applied after blooming control the disease.

Frogeye Leaf Spot

Frogeye leaf spot can occur on all soybean crops but crops in warm and humid regions are most susceptible. The main symptom is circular lesions on the foliage of the plant. As the number of lesions grows the leaves eventually wither and die. The disease can spread rapidly and often infects all the leaves of the plant. Yields from infected crops can be reduced by 15%. Fungicides applied at late flowering and the beginning of seed growth have been shown to protect against frogeye.

Rhizoctonia Aerial Blight

Rhizoctonia aerial blight attacks the soybean crop by infecting the plant during the flowering phase. It infects the tissue and produces lesions that eventually defoliate and rot the plant. Foliar fungicides have been shown to be effective at combating the disease.

Pod and Stem Blight

Pod and stem blight attack soybean plants during very wet seasons. They cause a fungal growth on the plant that eventually covers the entire plant. These fungi cause moldy, cracked, and shriveled seeds which in return produce low-quality oil and meal. Foliar fungicides applied between mid-flowing and late pod provide the best fungicidal control.

Spinach

Downy Mildew: Until the late 1950s the United States spinach industry was not considered highly viable. Cultivars were vulnerable to downy mildew, which caused sharp reductions in quality and in select cases complete crop loss. Control by fungicide use was not practical.

The spinach industry began to expand in 1947 when a strain of plant was introduced from Iran that was proven to be resistant to downy mildew. Therefore it was not until the late 1970s that a new strain of the disease made downy mildew a problem for U.S. growers.

The disease manifests as small yellow spots that quickly grow and rot the leaves. On the underside of the leaf blue mold spores grow and are spread to the rest of the spinach crop by wind. Epidemics occur quickly due to the fast growing nature of the fungus.

After the new strain was discovered and began to destroy spinach fields, testing began in the 1980s with fungicides to control the disease. Research showed that metalaxyl was effective, reducing yield loss from 43% to 1% in treated fields.

White Rust

White rust was discovered in Texas in 1937 and primarily occurs in spinach crops found in eastern U.S. production areas. It has not been reported outside of the U.S.

The main symptom of white rust is lesions on the leaf that grow and eventually release spores that cover and kill the entire leaf. Some cultivars of spinach have partial resistance to white rust but metalaxyl has been shown to be the best way to control the disease. It reduces yield loss from 50% in untreated plots to 1% in treated plots.

Sugarbeets

Powdery Mildew: Powdery mildew on sugarbeets can be found wherever the crop is grown in the U.S. It infects plants in the spring when the weather begins to get warm again. The spores land on the leaves and rapidly grow, forming a white mildew film on the foliage within a week of infection. This mildew penetrates the cell wall and begins to absorb nutrients from the leaves.

This eventually turns the leaves yellow and kills them. If it proceeds uncontrolled the infection will spread through an entire field within a month. Tests have shown that when left untreated powdery mildew can reduce yields by 27%. Control is best achieved through the use of sulfur, which has been found to increase yields by 38%.

Cercospora Leaf Spot

Cercospora leaf spot infects sugarbeets by overwintering in the soil from the year's previous crop. It can survive up to two years without a sugarbeet crop being planted. Once attached to the plant the fungus penetrates the tissue and begins to infect and kill nearby tissue.

As the disease progresses these individual rotted spots begin to come together to form massive areas of dead tissue. This causes severely infected leaves to eventually die. The plant responds by attempting to re-grow the leaf and using nutrients within the plant. This results in lowered sugar content and stunted growth. Losses due to the disease can be as high as 42%. Tetraconazole has been shown to provide the best control. Research demonstrated that treatments with tetraconazole increased the amount of sucrose extracted by 30%.

Sweet Corn

Northern Corn Leaf Blight

As there are many strains, northern corn leaf blight (NCLB) is considered one of the most serious and hard to control diseases of sweet corn. Its multiple strains mean it is nearly impossible to attempt any control of NCLB with plant resistance. The symptoms of the disease include long, thin dead areas on the leaf that slowly grow and coalesce. These lesions produce spores that can infect other plants. However much of the spread and severity of the disease is dependent upon weather conditions. Heavy rains, low temperatures, and relatively high humidity increase the number of life cycles of the fungus, producing additional spores. The most common control practice for NCLB is chlorothalonil, maneb, and mancozeb.

Southern Corn Leaf Blight

Southern corn leaf blight thrives in the southern United States because unlike NCLB it requires warm weather to develop. During favourable conditions the disease has been documented ruining a whole crop within days when left untreated. Treatment is available with the use of chlorothalonil, mancozeb, and maneb but even this is limited. In one study in Florida even when a regular fungicide plan was followed over 25% of the ears were lost.

Rust

Yield losses due to rust in corn can range from 20% to 50% of the crop depending upon when the infection strikes. If the disease attacks the corn early this will result in stunted growth of the ears and dried out kernels. However, this results in the lower end of crop yield loss.

The greater damage comes when the infection appears later in the season because, although yield remains the same, extensive cosmetic damage to the corn—pustules cover the leaves, husks, necks, and tassels of the plant—renders much of it becomes unmarketable. This can result in up to 50% loss of yield.

Biopesticide

Biopesticides, a contraction of 'biological pesticides', include several types of pest management intervention: through predatory, parasitic, or chemical relationships. The term has been associated historically with biological control - and by implication - the manipulation of living organisms. Regulatory positions can be influenced by public perceptions, thus:

- In the EU, biopesticides have been defined as "a form of pesticide based on micro-organisms or natural products".
- The US EPA states that they "include naturally occurring substances that control pests (biochemical pesticides), microorganisms that control pests (microbial pesticides), and pesticidal substances produced by plants containing added genetic material (plant-incorporated protectants) or PIPs".

They are typically created by growing and concentrating naturally occurring organisms and/or their metabolites including bacteria and other microbes, fungi, nematodes, proteins, *etc.* They are often considered to be important components of integrated pest management (IPM) programmes, and have received much practical attention as substitutes to synthetic chemical plant protection products (PPPs).

The Manual of Biocontrol Agents (2009: formerly the *Biopesticide Manual*) gives a review of the available biological insecticide (and other biology-based control) products.

Overview

Biopesticides fall into three major classes:

- Microbial pesticides which consist of bacteria, entomopathogenic fungi or viruses (and sometimes includes the metabolites that bacteria or fungi produce). Entomopathogenic nematodes are also often classed as microbial pesticides, even though they are multi-cellular.
- Biochemical pesticides are naturally occurring substances that control (or monitor in the case of pheromones) pests, but are relatively non-toxic to mammals.
- Plant-incorporated protectants (PIPs) have genetic material from other species incorporated into their genetic material (*i.e.* GM crops): still controversial in some (notably European) countries.

Biopesticides have usually no known function in photosynthesis, growth or other basic aspects of plant physiology; however, their biological activity against insect pests, nematodes, fungi and other organisms is well documented. Every plant species has developed a built-in unique chemical complex structure that protects it from pests.

The plant kingdom offers a diverse array of complex chemical structures and almost every imaginable biological activity. These biodegradable, economical and renewable alternatives are used especially under organic farming systems.

Examples

Bacillus thuringiensis, a bacterial disease of Lepidoptera, Coleoptera and Diptera, is a well-known insecticide example. Because it has little effect on other organisms, it is considered more environmentally friendly than synthetic pesticides.

The toxin from *B. thuringiensis* (Bt toxin) has been incorporated directly into plants through the use of genetic engineering.

Other microbial control agents include products based on:

- Entomopathogenic fungi (*e.g.Beauveria bassiana, Lecanicillium* spp., *Metarhizium* spp.),

- Plant disease control agents: include *Trichoderma* spp. and *Ampelomyces quisqualis* (a hyper-parasite of grape powdery mildew); *Bacillus subtilis* is also used to control plant pathogens.
- Beneficial nematodes attacking insect (*e.g. Steinernema feltiae*) or slug (*e.g. Phasmarhabditis hermaphrodita*) pests
- Entomopathogenic viruses (*e.g.. Cydia pomonella* granulovirus).
- Weeds and rodents have also been controlled with microbial agents.

Various naturally occurring materials, including fungal and plant extracts, have been described as biopesticides. Products in this category include:

- Insect pheromones and other semiochemicals
- Fermentation products such as Spinosad (a macro-cyclic lactone)
- *Chitosan:* a plant in the presence of this product will naturally induce systemic resistance (ISR) to allow the plant to defend itself against disease, pathogens and pests.
- Biopesticides may include natural plant-derived products, which include alkaloids, terpenoids, phenolics and other secondary chemicals. Certain vegetable oils such as canola oil are known to have pesticidal properties. Products based on extracts of plants such as garlic have now been registered in the EU and elsewhere.
- Naturally occurring minerals such as baking soda may also have pesticidal applications.

Applications

Biopesticides are typically microbial biological pest control agents that are applied in a manner similar to chemical pesticides. In order to implement these environmentally friendly pest control agents effectively, it can be important to pay attention to the way they are formulated and applied.

Biopesticides for use against crop diseases have already established themselves on a variety of crops. For example, biopesticides already play an important role in controlling downy mildew diseases.

Their benefits include: a 0-Day Pre-Harvest Interval, the ability to use under moderate to severe disease pressure, and the ability to use as a tank mix or in a rotational program with other registered fungicides.

Because some market studies estimate that as much as 20% of global fungicide sales are directed at downy mildew diseases, the integration of biofungicides into grape production has substantial benefits in terms of extending the useful life of other fungicides, especially those in the reduced-risk category.

A major growth area for biopesticides is in the area of seed treatments and soil amendments. Fungicidal and biofungicidal seed treatments are used to control soil borne fungal pathogens that cause seed rots, damping-off, root rot and seedling blights. They can also be used to control internal seed–borne fungal pathogens as well as fungal pathogens that are on the surface of the seed.

Many biofungicidal products also show capacities to stimulate plant host defenses and other physiological processes that can make treated crops more resistant to a variety of biotic and abiotic stresses.

Advantages

- Harmful residues not detected
- Can be cheaper than chemical pesticides when locally produced.
- Can be more effective than chemical pesticides in the long-term (as demonstrated, for example, by the LUBILOSA Programme)

Disadvantages

- *High specificity:* which may require an exact identification of the pest/pathogen and the use of multiple products to be used
- Often slow speed of action (thus making them unsuitable if a pest outbreak is an immediate threat to a crop)

- Often variable efficacy due to the influences of various biotic and abiotic factors (since biopesticides are usually living organisms, which bring about pest/pathogen control by multiplying within the target insect pest/ pathogen)
- Living organisms evolve and increase their resistance to biological, chemical, physical or any other form of control. If the target population is not exterminated or rendered incapable of reproduction, the surviving population can acquire a tolerance of whatever pressures are brought to bear, resulting in an evolutionary arms race.

Guidelines for Seed-Borne Diseases of Pulse Crops

Seed Quality and Seed-borne Diseases

Seed quality is very important in pulse production, as the costs of seed and seed treatments are a significant part of overall input costs.

"Quality seed" may have different meanings depending on whether the seed is being sold for food or feed, or being used for planting. If the seed is for planting, "quality" refers to seed that has a high germination, is intact, free from foreign materials and weed seeds, and has zero or low levels of seed-borne diseases. This article will focus on the importance of seed-borne diseases in quality pulse seed. Planting pulse seed that is free of ascochyta blight and other important seed-borne diseases is the primary means to limit the introduction of pathogens into a field, and prevent early establishment of disease.

Planting infected seed may result in a widespread distribution of disease within the crop, and an increased number of initial infection sites from which the disease can spread. As an example, consider the development of ascochyta blight in a chickpea crop. Since there is a high rate of seed-to-seedling transmission of this disease, even a small percentage of infected seed can result in significant seedling infection in the field. For a seed lot with 0.1 per cent ascochyta infection (one infected seed in 1,000 seeds) and a planting density of three to four plants/ft.2, 175 infected seedlings per acre could potentially result. This is a substantial amount of early infection for such an aggressive disease.

Importance of Seed Testing

It is recommended to have seed tested at an accredited laboratory to assess the levels of seed-borne diseases. Commercial laboratories can test pulse seed to determine the level of seed-borne fungi that cause ascochyta blight, anthracnose, botrytis seedling blight and grey mould, and sclerotinia. A parallel test for germination will indicate whether seed quality has been affected by such factors as immaturity, disease, mechanical injury and chemical damage. Testing for vigour may also be beneficial, as it serves as an indication of how seed will respond in less than ideal growing conditions.

Preferably, buy only certified seed with demonstrated good quality. However, keep in mind that certified seed, according to The Seeds Act (federal regulations), has only to meet standards for germination and purity, not for diseases. The motto is "buyer-beware" when purchasing pulse seed. Ask to see the lab certificate before purchasing seed.

Limitations of Disease Testing

A sample submitted to a lab is only a small representation of the larger seed lot, and its value is dependent on how the sample was collected. Due to the statistical nature of small samples, even a carefully selected representative sample may not quantify the exact disease level. For example, a zero per cent seed test result cannot guarantee that the entire seed lot is free of disease.

The number of seeds tested affects the accuracy of the test. Many pulse seed tests are done with a 400-seed sample, but a 1,000-seed sample is sometimes used for ascochyta blight of chickpea. By increasing the number of seeds to 1,000, the probability of detecting a small amount of infection is increased. For example, a 1,000-seed test can detect 0.1 per cent infection (one in 1,000), whereas a 400-seed sample test may have provided a zero per cent infection result. Thus, the 1,000-seed sample decreases the likelihood of getting a false negative result. The difference between zero per cent and 0.1 per cent infection can be important with a disease like ascochyta blight of chickpea, when weather conditions favour disease spread. The ability to detect low levels of infection and avoid false negative results is more important for chickpea than for lentil or field pea, and more important for seed growers than commercial growers.

Cleaning Seed

Diseased seed is often shrunken or discoloured, although this is not always the case. Severely ascochyta-infected seed may grade as No. 1 and still be carrying the fungus. Botrytis-infected seed is often shrunken. Cleaners and colour sorters can be used to remove some diseased or damaged seed to improve the seed quality for planting. Furthermore, it is recommended that producers have their seed cleaned and sized to allow for consistent flow in seeding equipment.

If requested, seed testing labs will clean seed to a standard level (based on screen size) before testing the lot for germination and disease. If the seed sample is very dirty, a disease test cannot be conducted without cleaning the seed first. As there is usually an extra charge to the client for cleaning to a suitable level, it is important to talk to the lab to confirm your expectations regarding cleaning.

Germination

Seed should be tested for germination to determine its suitability for planting. Germination can decrease in the bin over the winter, especially if the seed was immature or damaged at harvest. It is a good investment to re-test seed for germination in the spring, if quality was questionable in the fall. Increasing the seeding rate will compensate for low germination, but only to a certain extent.

If the reduced germination was a result of disease, an increased seeding rate can introduce more disease into the field. As well, do not use seed from a pulse crop that was treated with pre-harvest glyphosate. Chemically-damaged seed will show poor root development and may result in some seedlings dying and, in extreme cases, a complete crop failure.

Vigour

Some labs provide vigour testing, but there is still some debate about the best testing protocol and how to interpret results. Vigour is a measure of germination when seed is placed in less than ideal growing conditions, such as low temperatures. Some people believe that a vigour test is a valuable indicator of seed performance, as it mimics natural field conditions.

The smaller the gap between per cent germination and per cent vigour, the more sound the seed is believed to be. If there is a significant discrepancy between these two values, it is important to determine why the vigour was reduced, e.g. mechanical damage, a high proportion of green seed, herbicide damage, etc. Knowing the cause of the problem will help in the decision to plant the seed, or to source another seed lot with acceptable vigour.

The Value of Fungicide Seed Treatments

Fungicide seed treatments protect seed viability and inhibit diseases like seed rot and seed Fungicide seed treatments protect seed viability and inhibit diseases like seed rot and seedling blight. Seed treatments protect the seed in two ways: by controlling fungi present either on the seed surface or carried internally in the seed; and by controlling fungi present in the soil, or on crop residue in the soil. Treating seed ensures that the crop gets off to a good start. But keep in mind that seed treatments will not "cure" a poor seed lot that has high proportions of dead, damaged or infected seed.

The degree of control with seed treatment depends on five factors: (1) fungicide active ingredients, (2) rate of application, (3) seed- and soil-borne fungal diseases present, (4) environmental conditions, and (5) quality of seed coverage.

The latter point cannot be overlooked-full coverage of the product over the seed coat is essential to ensure protection.

If you are testing seed on-farm, equipment must be carefully calibrated to ensure proper mixing of the seed and chemical to provide adequate coverage.

This becomes even more important for fungicides with contact activity. Large-sized pulse seed may be easily damaged during treatment, so reduce velocity within the equipment and during augering if possible. The rate of application listed on the product label must be adhered to, because over-treatment may injure the seed, and under treatment may not provide adequate disease control.

In general, seed treatments may have either systemic or contact modes of action. Controlling fungi that are carried within the seed requires a systemic product (i.e. smut in barley), whereas contact or protectant products are adequate for surface-borne or soil-borne fungi.

Systemic seed treatments are diluted quite quickly within the plant once the seed germinates and is actively growing. Some treatments will protect a young seedling against early leaf disease or root rot infection, but in most cases, seed treatments are no longer effective after seedling emergence.

Note: If seed-applied Rhizobium inoculants are being used as well, it is recommended to first apply the fungicide seed treatment, allow it to dry on the seed, then add the inoculant as close to planting as possible. Contact your inoculant and crop protection provider for product-specific compatibility information.

Guidelines for Tolerance Levels for Planting

There are no reseach-based recommendations for what levels of seed-borne infection in pulse seed are tolerable for planting. The critical level will depend on moisture and temperature conditions after planting and will, therefore, vary from year to year.

The decision whether to plant infected seed will also depend on a variety of factors affecting risk management.

Factors to consider when planting infected seed include:

- The cost and availability of disease-free seed with good germination.
- The cost and availability of registered seed treatments.
- The weather conditions and disease pressure typical of the region/soil climatic zone.
- The type of pulse crop.
- The type of disease present in the seed and proximity to other sources of disease inoculum.
- Plans for infield scouting; availability and costs of foliar fungicides.

Despite the above variables, the following guidelines were developed to give pulse producers some assistance when making decisions about seed-borne diseases.

These guidelines are based partly on knowledge of the biology of different diseases, partly on published research, and partly on crop insurance considerations. However, because of the biological, agronomic, environmental and economic variables mentioned, the guidelines should be treated as rules of thumb.

A Review on Biological Control of Fungal Plant Pathogens Using Microbial Antagonists

The objective of this study was to review the published research works on biological control of fungal plant diseases during past 50 years. Fungal plant pathogens are among the most important factors that cause serious losses to agricultural products every year.

Biological control of plant diseases including fungal pathogens has been considered a viable alternative method to chemical control. In plant pathology, the term biocontrol applies to the use of microbial antagonists to suppress diseases. Throughout their lifecycle, plants and pathogens interact with a wide variety of organisms.

These interactions can significantly affect plant health in various ways. Different mode of actions of biocontrol-active microorganisms in controlling fungal plant diseases include hyperparasitism, predation, antibiosis, cross protection, competition for site and nutrient and induced resistance. Successful application of biological control strategies requires more knowledge-intensive management.

Various methods for application of biocontrol agents include: application directly to the infection court at a high population level to swamp the pathogen, application at one place in which biocontrol microorganisms are applied at one place (each crop year) but at lower populations which then multiply and spread to other plant parts and give protection against pathogens and one time or occasional application that maintain pathogen populations below threshold levels.

Commercial use and application of biological disease control have been slow mainly due to their variable performances under different environmental conditions in the field. To overcome this problem and in order to take the biocontrol technology to the field and improve the commercialization of biocontrol, it is important to develop new formulations of biocontrol microorganisms with higher degree of stability and survival.

Majority of biocontrol products are applied against seed borne and soil borne fungal pathogens, including the causal agents of seed rot, damping-off and root rot diseases. These products are mostly used as seed treatment and have been effective in protecting several major crops such as wheat, rice, corn, sugar beet and cotton against fungal pathogens.

However, in some cases, biocontrol microorganisms have also been tested as spray application on foliar diseases, including powdery mildew, downy mildew, blights and leaf spots. A few post harvest fungal diseases have also been controlled by the use of antagonistic fungi and bacteria.

Biocontrol microorganisms are also being used as the form of composts in some plants. Research data and observations in nurseries have shown that addition of composted organic matter to potting mixes results in suppression of soil borne diseases.

A significant improvement have been made in different aspects of biological control of fungal plant diseases, but this area still need much more development and investigations to solve the existing problems.

In order to have more effective biological control strategies in the future, it is critical to carry out more research studies on some less developed aspects of biocontrol, including development of novel formulations, understanding the impact of environmental factors on biocontrol agents, mass production of biocontrol microorganisms and the use of biotechnology and nano-technology in improvement of biocontrol mechanisms and strategies.

Future outlooks of biocontrol of plant diseases is bright and promising and with the growing demand for biocontrol products among the growers, it is possible to use the biological control as an effective strategy to manage plant diseases, increase yield, protect the environment and biological resources and approach a sustainable agricultural system.

Plant pests (harmful insects, parasitic weeds and pathogens) are among the most important biotic agents causing serious losses and damages to agricultural products. Plant pests need to be controlled to ensure food, feed and fiber production quantitatively and qualitatively. A number of different strategies are currently being employed to manage and control plant pests (Agrios, 1988; Baker, 1987; Cook, 1993; Bargabus *et al.*, 2002, 2004; Benhamou, 2004; Chisholm *et al.*, 2006; Heydari, 2007; Heydari *et al.*, 2007; Islam *et al.*, 2005; Kloepper*et al.*, 2004).

Beyond good agronomic and cultural practices, growers often rely heavily on chemical pesticide application (Agrios, 1988; Baker, 1987). However, the environmental pollution caused by excessive use of agrochemicals has led to considerable changes in people's attitudes towards the use of pesticides in agriculture.

Today, there are strict regulations on chemical pesticide use and there is political pressure to remove the most hazardous chemicals from the market. In addition to the above-mentioned issues, the spread of plant diseases in natural ecosystems may preclude successful application of chemicals, because of the scale to which such applications might have to be applied. Consequently, some pest management researchers have focused their efforts on developing alternative inputs to synthetic chemicals for controlling pests and diseases (Baker, 1987; Cook, 1993).

Plant diseases are mostly controlled by the use of chemical pesticides and in some cases by cultural practices (Agrios, 1988; Cook, 1993). However, the widespread use of chemicals in agriculture has been a subject of public concern and scrutiny due to the potential harmful effects on the environment, their undesirable effects on non-target organisms and possible carcinogenicity of some chemicals (Agrios, 1988; Cook, 1993; Heydari, 2007; Heydari *et al.*, 2007).

Other problems include development of resistant races of pathogens, a gradual elimination and phasing out of some available pesticides and the reluctance of some chemical companies to develop and test new chemicals due to the problems with registration process and cost (Cook, 1993). The need for the development of non-chemical alternative methods to control plant diseases is therefore clear.

Biological control of plant diseases has been considered a viable alternative method to manage plant diseases (Cook, 1993). Biological control is the inhibition of growth, infection or reproduction of one organism using another organism (Cook, 1993; Baker, 1987). Biocontrol is environmentally safe and in some cases is the only option available to protect plants against pathogens (Cook, 1993).

Biological control employs natural enemies of pests or pathogens to eradicate or control their population. This can involve the introduction of exotic species, or it can be a matter of harnessing whatever form of biological control exists naturally in the ecosystem.

The induction of plant resistance using non-pathogenic or incompatible microorganisms is also a form of biological control (Cook, 1993; Schouten *et al.*, 2004). Fungal plant diseases are considered the most important microbial agents causing serious losses in the agriculture annually (Agrios, 1988). Some fungal diseases that have successfully been controlled using biological agents are pathogens of pruning wounds and other cut surfaces, diseases of leaves and flowers, such as powdery mildew, diseases of fruits and vegetables, such as Botrytis and fungal pathogens in the soil (Agrios, 1988; Baker, 1987; Cook, 1993; Heydari, 2007; Heydari *et al.*, 2004, 2007; Heydari and Misaghi, 1998, 1999, 2003).

A variety of biological controls are available for use, but further development and effective adoption will require a greater understanding of the complex interactions among plants, people and the environment. To that end, the objectives of this review chapter is to present an advanced survey of the nature and practice of biological control as it is applied to the suppression of plant diseases. In this review, different aspects of biological control of fungal plant diseases including definitions, modes of action, application strategies, current status and future development and outlooks will be discussed.

Terminology

The term biological control and its abbreviated synonym biocontrol have been used in different fields of biology, most notably entomology and plant pathology. In plant pathology, the term applies to the use of microbial antagonists to suppress diseases as well as the use of host-specific pathogens to control weed populations (Cook, 1993).

In both fields, the organism that suppresses the pest or pathogen is referred to as the Biological Control Agent (BCA). More broadly, the term biological control also has been applied to the use of the natural products extracted or fermented from various sources (Cook, 1993).

These formulations may be very simple mixtures of natural ingredients with specific activities or complex mixtures with multiple effects on the host as well as the target pest or pathogen. While such inputs may mimic the activities of living organisms, non-living inputs should more properly be referred to as biopesticides or biofertilizers, depending on the primary benefit provided to the host plant (Cook, 1993).

The various definitions offered in the scientific literature have sometimes caused confusion and controversy. For example, members of the United States National Research Council took into account modern biotechnological developments and referred to biological control as the use of natural or modified organisms, genes, or gene products, to reduce the effects of undesirable organisms and to favour desirable organisms such as crops, beneficial insects and microorganisms, but this definition spurred much subsequent debates and it was frequently considered too broad by many scientists who worked in the field.

Published definitions of biocontrol differ depending on the target of suppression; number, type and source of biological agents and the degree and timing of human intervention (Cook, 1993). Most broadly, biological control is the suppression of damaging activities of one organism by one or more other organisms, often referred to as natural enemies.

With regards to plant diseases, suppression can be accomplished in many ways. If grower's activities are considered relevant, cultural practices such as the use of rotations and planting of disease resistant cultivars (whether naturally selected or genetically engineered) would be included in the definition (Cook, 1993).

Because the plant host responds to numerous biological factors, pathogenic and non-pathogenic, induced host resistance might be considered a form of biological control (Cook, 1993).

More narrowly, biological control refers to the purposeful utilization of introduced or resident living organisms, other than disease resistant host plants, to suppress the activities and populations of one or more plant pathogens.

This may involve the use of microbial inoculants to suppress a single type or class of plant diseases. This may also involve managing soils to promote the combined activities of native soil and plant-associated organisms that contribute to general suppression (Cook, 1993). Most narrowly, biological control refers to the suppression of a single pathogen by a single antagonist, in a single cropping system. Most specialists in the field would concur with one of the narrower definitions presented above.

Interactions Between Plants and Beneficial Microbes

Throughout their lifecycle, plants and pathogens interact with a wide variety of organisms. These interactions can significantly affect plant health in various ways (Agrios, 1988; Bull *et al.*, 2002; Katska, 1994; Chisholm *et al.*, 2006; Fitter and Garbaye, 1994; McSpadden-Gardener and Weller, 2001).

In order to understand the mechanisms of biological control, it is helpful to appreciate the different ways that organisms interact. Note too, in order to interact, organisms must have some form of direct or indirect contact. The types of interactions between plants and microorganisms have been referred to as mutualism, protocooperation, commensalisms, neutralism, competition, amensalism, parasitism and predation (Bankhead *et al.*, 2004; Bull *et al.*, 2002; Katska, 1994; Chisholm *et al.*, 2006; Fitter and Garbaye, 1994; Hoitink and Boehm, 1999). While the terminology has been developed for macroecology, examples of all of these types of interactions can be found in the natural world at both the macroscopic and microscopic level. And, because the development of plant diseases involves both plants and microbes, the interactions that lead to biological control take place at multiple levels of scale (Bull *et al.*, 2002; Katska, 1994; Chisholm *et al.*, 2006; Fitter and Garbaye, 1994).

From the plant's point of view, biological control may be considered a positive result arising from different specific and non-specific interactions (Cook, 1993; Weller *et al.*, 2002). We can begin to classify and functionally delineate the diverse components of ecosystems that contribute to biological control.

Mutualism is an association among several species where all of them are benefited from this association (Biermann and Linderman, 1983; Bull *et al.*, 2002; Katska, 1994; Chisholm *et al.*, 2006; Duchesne, 1994; Fitter and Garbaye, 1994; Garcia-Garrido and Ocampo, 1989; Kerry, 2000).

Sometimes, it can be an obligatory relation involving close physical and biochemical contact between two organisms, such as those between plants and mycorrhizal fungi (Bull *et al.*, 2002; Katska, 1994; Chisholm *et al.*, 2006; Fitter and Garbaye, 1994). However, they are generally facultative and opportunistic.

For example, Rhizobium bacteria reproduce either in the soil or, to a much greater degree, through their mutualistic association with legume plants.

These types of mutualism can contribute to biological control, by providing plant with improved nutrition and/or by stimulating host defense mechanism and ability (Chisholm *et al.*, 2006; Fitter and Garbaye, 1994). Many of the microorganisms isolated and classified as biocontrol agents (BCA) can be considered facultative mutualists, because host and disease suppression by them will vary depending on the prevailing environmental conditions (Cook, 1993).

Commensalism is also a symbiotic interaction between two living organisms, where one organism benefits and the other is neither harmed nor benefited (Fitter and Garbaye, 1994). Most plant-associated microorganisms are assumed to be commensals with regards to the host plant, because their presence, individually or in total, rarely results in positive or negative consequences to the plant (Katska, 1994; Chisholm *et al.*, 2006). While the presence of these microorganisms may present a variety of challenges to an infecting pathogen, their absence decreases pathogen infection or disease severity and is indicative of commensal interactions (Cook, 1993).

Biological interactions in which the population density of one species has absolutely no effect on the other are called neutralism (Berg *et al.*, 2005; Chisholm *et al.*, 2006). Related to biological control, an inability to associate the population dynamics of pathogen with that of another organism would indicate neutralism (Chisholm *et al.*, 2006). In contrast, antagonism between organisms results in a negative outcome for one or both. Competition within and between species caused a decreased growth, activity, and/or fecundity of the interacting organisms (Cook, 1993). Biocontrol can occur when non-pathogens compete with pathogens for nutrients and sites in host plant. Direct interactions that benefit one population at the expense of another also affect our understanding of biological control (Cook, 1993).

Parasitism is also a symbiotic relation in which two organisms coexist over a prolonged period of time (Cook, 1993; Chisholm *et al.*, 2006; Lo *et al.*, 1997). In this type of interaction, one organism, usually the physically smaller (parasite) benefits and the other (host) is harmed.

The activities of various hyperparasites, for example those agents that parasitize plant pathogens, can result in biocontrol (Lo *et al.*, 1997). Another interesting contribution to biocontrol is when host infection and parasitism by relatively avirulent pathogens may lead to biocontrol of more virulent pathogens through the stimulation of host defense systems (Cook, 1993). Finally predation refers to the hunting and killing of one organism by another for consumption and sustenance.

While the term predator typically refers to animals that feed at higher trophic levels in the macroscopic world, it has also been applied to the actions of microorganisms such as protists and mesofauna, e.g. fungal feeding nematodes and microarthropods, that consume pathogen biomass for sustenance (Cook, 1993).

Biological control can result in various forms of these types of interactions, depending on the environmental conditions within which they occur. Significant biological control, as was described above, generally arises from manipulating mutualisms between microorganisms and their plant hosts or from manipulating antagonisms between microbes and pathogens (Bull *et al.*, 2002; Katska, 1994; Chisholm *et al.*, 2006; Fitter and Garbaye, 1994).

Mechanisms of Biological Control

Since biological control is a result of many different types of interactions among microorganisms, scientists have concentrated on characterization of mechanisms occurring in different experimental situations (Audenaert *et al.*, 2002; De

Meyer and Hofte, 1997; Elad and Baker, 1985; Heydari *et al.*, 1997; Homma *et al.*, 1989; Howell *et al.*, 1988; Islam *et al.*, 2005; Meziane *et al.*, 2005; Ryu *et al.*, 2004; Van Dijk and Nelson, 2000). In all cases, pathogens are antagonized by the presence and activities of other microorganisms that they encounter.

Direct antagonism results from physical contact and/or a high-degree of selectivity for the pathogen by the mechanism(s) expressed by the biocontrol active microorganisms. In this type of interaction, Hyperparasitism by obligate parasites of a plant pathogen would be considered the most direct type of mechanism because the activities of no other organism would be required to exert a suppressive effect (Harman *et al.*, 2004; Linderman, 1994). In contrast, indirect antagonism is resulted from the activities that do not involve targeting a pathogen by a biocontrol active microorganism.

Improvement and stimulation of plant host defense mechanism by non-pathogenic microorganisms is the most indirect form of antagonism (Kloepper *et al.*, 1980; Lafontaine and Benhamo, 1996; Leeman *et al.*, 1995; Maurhofer *et al.*, 1994; Silva *et al.*, 2004). While many studies have concentrated on the establishment of the importance of specific mechanisms of biocontrol to particular pathosystems, all of the mechanisms described below are likely to be operating to some extent in all natural and managed ecosystems.

The most effective biocontrol active microorganisms studied appear to antagonize plant pathogens employing several modes of actions (Cook, 1993). For example, pseudomonads known to produce the antibiotic 2, 4-diacetylphloroglucinol (DAPG) may also induce host defenses (Kloepper *et al.*, 1980; Lafontaine and Benhamou, 1996; Leeman *et al.*, 1995; Maurhofer *et al.*, 1994; Silva *et al.*, 2004). Additionally, DAPG-producers bacterial antagonists can aggressively colonize roots, a trait that might further contribute to their ability to suppress pathogen activity in the rhizosphere of plant through competition for organic nutrients. However, the most important modes of actions of biocontrol active microorganisms are as follows:

Mycoparasitism: In Hyperparasitism, the pathogen is directly attacked by a specific biocontrol agent (BCA) that kills it or its propagules. Four major groups of hyperparasites have generally been identified which include hypoviruses, facultative parasites, obligate bacterial pathogens and predators.

An example of hypoparasites is the virus that infects *Cryphonectria parasitica*, the fungal causal agent of chestnut blight, which causes hypovirulence, a reduction in pathogenicity of the pathogen. This phenomenon has resulted in the control of chestnut blight in many places (Milgroom and Cortesi, 2004). However, the interaction of virus, fungus, tree and environment determines the success or failure of hypovirulence.

In addition to hypoviruses several fungal hypoparasites have also been identified including those that attack sclerotia (e.g., *Coniothyrium minitans*) or others that attack fungal hyphae (e.g. *Pythium oligandrum*).

In some cases, a single fungal pathogen can be attacked by multiple hyperparasites. For example, *Acremonium alternatum*, *Acrodontium crateriforme*, *Ampelomyces quisqualis*, *Cladosporium oxysporum* and *Gliocladium virens* are just a few of the fungi that have the capacity to parasitize powdery mildew pathogens (Milgroom and Cortesi, 2004).

In contrast to hyperparasitism, microbial predation is more general, non-specific and generally provides less predictable levels of disease control.

Some biocontrol agents exhibit predatory behaviour under nutrient-limited conditions. Such as *Trichoderma*, a fungal antagonist that produces a range of enzymes that are directed against cell walls of pathogenic fungi.

However, when fresh bark is used in composts, *Trichoderma* sp. does not directly attack the plant pathogen, *Rhizoctonia solani*. But, in decomposing bark, the concentration of readily available cellulose decreases and this activates the chitinase genes of *Trichoderma* sp. Which, in turn, produce chitinase to parasitize *R. solani* (Benhamou and Chet, 1997).

Antibiosis: Many microbes produce and secrete one or more compounds with antibiotic activity (Homma *et al.*, 1989; Howell and Stipanovic, 1980; Islam *et al.*, 2005; Leclére *et al.*, 2005; Shahraki *et al.*, 2009; Shanahan *et al.*, 1992; Thomashow *et al.*, 1990; Thomashow and Weller, 1988). In a general definition antibiotics are microbial toxins that can, at low concentrations, poison or kill other microorganisms. It has been shown that some antibiotics produced by microorganisms are particularly effective against plant pathogens and the diseases they cause (Homma *et al.*, 1989; Howell and Stipanovic, 1980; Islam *et al.*, 2005; Shanahan *et al.*, 1992; Thomashow *et al.*, 1990, 2002; Thomashow and Weller, 1988). In all cases, the antibiotics have been shown to be particularly effective at suppressing growth of the target pathogen *in vitro* and/or *in situ* conditions.

An effective antibiotic must be produced in sufficient quantities (dose) near the pathogen. *In situ* production of antibiotics by several different biocontrol agents has been studied (Thomashow *et al.*, 1990). While several procedures have been developed to ascertain when and where biocontrol agents may produce antibiotics detecting expression in the infection court is difficult because of the heterogenous distribution of plant-associated microbes and the potential sites of infection (Thomashow *et al.*, 1990).

However, in some cases, the relative importance of antibiotic production by biocontrol bacteria has been demonstrated. For example, mutant strains incapable of producing phenazines (Thomashow and Weller, 1988) or phloroglucinols (Keel *et al.*, 1989) have been shown to be equally capable of colonizing the rhizosphere, but much less capable of suppressing soil borne root diseases than the corresponding wild-type and complemented mutant strains. Many biocontrol strains have been shown to produce multiple antibiotics which can suppress one or more pathogens (Homma *et al.*, 1989; Howell and Stipanovic, 1980; Islam *et al.*, 2005; Shanahan *et al.*, 1992; Thomashow *et al.*, 1990; Thomashow and Weller, 1988). The ability of production of several antibiotics probably results in suppression of diverse microbial competitors and plant pathogens.

Metabolite Production: Many biocontrol active microorganisms produce other metabolites that can interfere with pathogen growth and activities. Lytic enzymes are among these metabolites that can break down polymeric compounds, including chitin, proteins, cellulose, hemicellulose and DNA (Anderson *et al.*, 2004; Howell *et al.*, 1988; Loper and Buyer, 1991; Ordentlich *et al.*, 1988; Press *et al.*, 2001; Wilhite *et al.*, 2001). Studies have shown that some of these metabolites can sometimes directly result in the suppression of plant pathogens.

For example, control of *Sclerotium rolfsii* by *Serratia marcescens* appeared to be mediated by chitinase expression (Ordentlich *et al.*, 1988). It seems more likely that antagonistic activities of these metabolites are indicative of the need to degrade complex polymers in order to obtain carbon nutrition.

Micro-organisms that show a preference in colonizing and suppression of plant pathogens might be classified as biocontrol agents. For example, Lysobacter and Myxobacteria that produce lytic enzymes have been shown to be effective against some plant pathogenic fungi (Bull *et al.*, 2002).

Studies have shown that some products of lytic enzyme activity may have indirect efficacy against plant pathogens. For example, oligosaccharides derived from fungal cell walls have been shown to induce plant host defenses (Howell *et al.*, 1988). It is believed that the effectiveness of the above compounds against plant pathogens is dependent on the composition and carbon and nitrogen sources of the soil and rhizosphere.

For example, in post-harvest disease control, addition of chitosan which is a non-toxic and biodegradable polymer of beta-1, 4-glucosamine produced from chitin by alkaline deacylation stimulated microbial degradation of pathogens (Benhamou, 2004). Amendment of plant growth substratum with chitosan suppressed the root rot caused by *Fusarium oxysporum* f. sp. radicis-lycopersici in tomato (Lafontaine and Benhamou, 1996).

In addition to the above-mentioned metabolites, other microbial byproducts may also play important roles in plant disease biocontrol (Phillips *et al.*, 2004). For example, Hydrogen cyanide (HCN) effectively blocks the cytochrome oxidase pathway and is highly toxic to all aerobic microorganisms at picomolar concentrations (Ramette *et al.*, 2003). The production of HCN by certain fluorescent pseudomonads is believed to be effective against plant pathogens.

Results of some research studies in this regard have shown that *P. fluorescens* CHA0, an antagonistic bacterium, produces antibiotics including siderophores and HCN, but suppression of black rot of tobacco caused by *Thielaviopsis basicola* appeared to be due primarily to HCN production. In another study Howell *et al.* (1988) reported that volatile compounds such as ammonia produced by *Enterobacter cloacae* were involved in the suppression of cotton seedling damping-off caused by *Pythium ultimum*.

Competition: The nutrient sources in the soil and rhizosphere are frequently not sufficient for microorganisms. For a successful colonization of phytosphere and rhizosphere a microbe must effectively compete for the available nutrients (Elad and Baker, 1985; Keel *et al.*, 1989; Loper and Buyer, 1991). On plant surfaces, host-supplied nutrients include exudates, leachates, or senesced tissue.

In addition to these, nutrients can also be obtained from waste products of other organisms such as insects and the soil. This is a general believe that competition between pathogens and non-pathogens for nutrient resources is an important issue in biocontrol (Elad and Baker, 1985; Keel *et al.*, 1989; Loper and Buyer, 1991).

Ii is also believed that competition for nutrients is more critical for soil borne pathogens, including *Fusarium* and *Pythium* species that infect through mycelial contact than foliar pathogens that germinate directly on plant surfaces and infect through appressoria and infection pegs (Elad and Baker, 1985; Keel *et al.*, 1989; Loper and Buyer, 1991).

Results of a study by Anderson *et al.* (1988) revealed that production of a particular plant glycoprotein called agglutinin was correlated with potential *of Pseudomonas putida* to colonize the root system. *P. putida* mutants deficient in this ability exhibited reduced capacity to colonize the rhizosphere and a corresponding reduction in Fusarium wilt suppression in cucumber (Tari and Anderson, 1988).

It has been shown that non-pathogenic plant-associated microrganisms generally protect the plant by rapid colonization and thereby exhausting the limited available substrates so that none are available for pathogens to grow. For example, effective catabolism of nutrients in the spermosphere has been identified as a mechanism contributing to the suppression of *Pythium ultimum* by *Enterobacter cloacae* (Van Dijk and Nelson, 2000; Kageyama and Nelson, 2003). At the same time, these microbes produce metabolites that are effective in suppression of pathogens. These microbes colonize the sites where water and carbon-containing nutrients are most readily available, such as exit points of secondary roots, damaged epidermal cells and nectaries and utilize the root mucilage.

Competition for rare but essential micronutrients, such as iron, has also been shown to be important in biological disease control. Iron is extremely limited in the rhizosphere, depending on soil pH. In highly oxidized and aerated soil, iron is present in ferric form (Kageyama and Nelson, 2003; Shahraki *et al.*, 2009), which is insoluble in water and the concentration may be extremely low. This very low concentration can not support the growth of microorganisms. To survive in such environment, organisms were found to secrete iron-binding ligands called Siderophores having high ability to obtain iron from the micro-organisms (Shahraki *et al.*, 2009). Almost all microorganisms produce siderophores, of either the catechol type or hydroxamate type (Kageyama and Nelson, 2003).

A direct correlation was established in vitro between siderophore synthesis in fluorescent pseudomonads and their capacity to inhibit germination of chlamydospores of *F. oxysporum* (Elad and Baker, 1985). It was shown that mutants incapable of producing some siderophores, such as pyoverdine, were reduced in their capacity to suppress different plant pathogens (Keel *et al.*, 1989; Loper and Buyer, 1991). The increased efficiency in iron uptake of the commensal microorganisms is thought to be a critical factor in their root colonization ability which is a major factor in biocontrol performance of bacterial antagonists.

Induction of Resistance: Plants actively respond to a variety of environmental stimulating factors, including gravity, light, temperature, physical stress, water and nutrient availability and chemicals produced by soil and plant associated microorganisms (Audenaert *et al.*, 2002; De Meyer and Hofte, 1997; Kloepper *et al.*, 1980; Leeman *et al.*, 1995; Moyne *et al.*, 2000; Vallad and Goodman, 2004; Van Loon *et al.*, 1998; Van Peer and Schippers, 1992; Van Wees *et al.*, 1997). Such stimuli can either induce or condition plant host defenses through biochemical changes that enhance resistance against subsequent infection by a variety of pathogens. Induction of host defenses can be local and/or systemic in nature, depending on the type, source and amount of stimulation agents (Audenaert *et al.*, 2002; De Meyer and Hofte, 1997; Kloepper *et al.*, 1980; Leeman *et al.*, 1995; Moyne *et al.*, 2000; Vallad and Goodman, 2004; Van Loon *et al.*, 1998; Van Peer and Schippers, 1992; Van Wees *et al.*, 1997).

Recently, plant pathologists have begun to characterize the determinants and pathways of induced resistance stimulated by biological control agents and other non-pathogenic micro-organisms (Audenaert *et al.*, 2002; Moyne *et al.*, 2000; Vallad and Goodman, 2004).

The first pathway called Systemic Acquired Resistance (SAR), is mediated by Salicylic Acid (SA), a chemical compound which is usually produced after pathogen infection and typically leads to the expression of Pathogenesis-related (PR) proteins (Vallad and Goodman, 2004). These PR proteins include a variety of enzymes some of which may act directly to lyse invading cells, reinforce cell wall boundaries to resist infections, or induce localized cell death (Vallad and Goodman, 2004).

Second pathway, called Induced Systemic Resistance (ISR), is mediated by Jasmonic Acid (JA) and/or ethylene, which are produced following applications of some nonpathogenic rhizobacteria (Audenaert *et al.*, 2002; De Meyer and Hofte, 1997; Kloepper *et al.*, 1980; Leeman *et al.*, 1995; Moyne *et al.*, 2000; Van Loon *et al.*, 1998; Van Peer and Schippers, 1992; Van Wees *et al.*, 1997). Interestingly, the SA- and JA- dependent defense pathways can be mutually antagonistic and some bacterial pathogens take advantage of this to overcome the SAR.

For example, pathogenic strains of *Pseudomonas syringae* produce coronatine, which is similar to JA, to overcome the SA-mediated pathway (Vallad and Goodman, 2004). Since the various host-resistance pathways can be activated to variable degrees by different microorganisms and insect feeding, it is therefore possible that multiple stimuli are constantly being received and processed by the plant. Thus, the magnitude and duration of host defense induction will likely vary over time. Only if induction can be controlled, i.e., by overwhelming or synergistically interacting with endogenous signals, will host resistance be increased (Audenaert *et al.*, 2002; De Meyer and Hofte, 1997; Kloepper *et al.*, 1980; Leeman *et al.*, 1995; Moyne *et al.*, 2000).

Some strains of root-colonizing microorganisms have been identified as potential elicitors of plant host defenses. For example, some biocontrol active strains of *Pseudomonas* sp. and *Trichoderma* sp. are known to strongly induce plant host defenses (Haas and Defago, 2005; Harman *et al.*, 2004).

In other instances, inoculation with Plant Growth Promoting Rhizobacteria (PGPR) have been shown to be effective in controlling multiple diseases caused by different fungal pathogens, including anthracnose (*Colletotrichum lagenarium*). A number of chemical elicitors of SAR and ISR such as salicylic acid, siderophore, lipopolysaccharides and 2, 3-butanediol may be produced by the PGPR strains upon inoculation (Ryu *et al.*, 2004; Van Loon *et al.*, 1998).

A substantial number of microbial products have been reported to elicit host defenses, indicating that host defenses are likely stimulated continually during the plant's lifecycle (Ryu *et al.*, 2004; Van Loon *et al.*, 1998). These inducers include lipopolysaccharides and flagellin from Gram-negative bacteria; cold shock proteins of diverse bacteria; transglutaminase, elicitins and a-glucans in Oomycetes; invertase in yeast; chitin and ergosterol in all fungi; and xylanase in Trichoderma (Ryu *et al.*, 2004). These findings indicate that plants would detect the composition of their plant-associated microbial communities and respond to changes in the quantity, quality and localization of many different signals.

The importance of such interactions is indicated by the fact that further induction of host resistance pathways, by chemical and microbiological inducers, is not always effective in improving plant health or productivity in the field (Vallad and Goodman, 2004).

Methods of Application of Antagonists

Overall application: Successful application of biological control strategies requires more knowledge-intensive management (Baker, 1987; Cook, 1993; Heydari *et al.*, 2004; Shah-Smith and Burns, 1997).

Understanding when and where biological control of plant pathogens can be profitable, requires an appreciation of its place within integrated pest management systems (Cook, 1993; Heydari *et al.*, 2004; Shah-Smith and Burns, 1997).

In general, the foundation of a sound pest and disease management program in an annual cropping system begins with cultural practices that alter the farm landscape to promote crop health (Cook, 1993; Heydari *et al.*, 2004; Shah-Smith and Burns, 1997). These include crop rotations that limit the availability of host material used by plant pathogens (Cook, 1993). Proper use of tillage can disrupt pathogen life cycles and prepare seed beds of optimal moisture and bulk density.

Careful management of soil fertility and moisture can also limit plant diseases by minimizing plant stress (Cook, 1993). In nurseries and greenhouses environmental control can be more tightly regulated in terms of temperature, light, moisture and soil composition, but the design of such systems cannot wholly eliminate disease problems (Paulitz and Belanger, 2001).

The second layer of defense against pests consists of the quality of crop germplasm. Breeding for pathogen resistance including fungal pathogens contributes substantially to crop success in most regions (Cook, 1993). Newer technologies that directly incorporate genes into crop genomes, commonly referred to as genetic modification or genetic engineering, are bringing new traits into crop.

Other technologies, such as seed washing, testing for pathogens and treatments are also used to keep germplasm pathogen-free. In perennial cropping systems, such as orchards and forests, germplasm quality may be more important than cultural practices, because rotation and tillage cannot be used as regularly (Agrios, 1988; Cook, 1993). Upon these two layers, growers can further reduce pathogen pressure by considering both biological and chemical inputs.

Biologically based inputs such as microbial fungicides can be used to interfere with pathogen activities. Registered biofungicides are generally labelled with short reentry intervals and pre-harvest intervals, giving greater flexibility to growers who need to balance their operational requirements and disease management goals.

When living microorganisms are introduced, they may also augment natural beneficial populations to further reduce the damage caused by targeted pathogens (Cook, 1993; Heydari *et al.*, 2004; Shah-Smith and Burns, 1997).

Applying to the Infection Site: Application directly to the infection court at a high population level to swamp the pathogen (inundate application), seed coating and treatment with antagonistic fungi and bacteria, e.g., *Trichoderma harzianum* and *Psudomonas fluorescens* (Cook, 1993; Heydari and Misaghi, 2003; Heydari *et al.*, 2004), antagonists applied to fruit for protection in storage, e.g., *Pseudomonas fluorescens* (De Capdeville *et al.*, 2002; El-Ghaouth *et al.*, 2000; Janisiewicz and Korsten, 2002; Janisiewicz and Peterson, 2004) and application to soil at the site of seed placement (Heydari and Misaghi, 2003). These types of applications are the most commonly used procedures which have resulted in the successful control of several fungal plant pathogens.

One Place Application: in this procedure, biocontrol microorganisms are applied at one place (each crop year), but at lower populations which then multiply and spread to other plant parts and give protection (augmentative application) against fungal pathogens. An Example of this method is Plant Growth Promoting Rhizobacteria (PGPR) and atoxigenic *Aspergillus flavus* on wheat seed scattered on the soil to spread to cotton flowers where they displace aflatoxin producing strains of *A. flavus* and fungal antagonists added to soil (Islam *et al.*, 2005; Kloepper *et al.*, 2004).

Occasional Application: One time or occasional application maintains pathogen populations below threshold levels. In theory, parasites of the pathogen, or hypovirulent (disease carrying) strains of the pathogen, might be used and not require yearly repetition (e.g., hypovirulent strains of the chestnut blight pathogen) in which host plant is inoculated with attenuated strains of pathogenic that protects the host plant against the virulent strains of pathogen (Milgroom and Cortesi, 2004).

Biocontrol of Different Fungal Pathogens

Microorganisms naturally present in the plants ecosystem will help reduce disease potential or disease damage, but only if they are allowed to grow vigorously (Cook, 1993). They accomplish these tasks by competing with the pathogens for food sources, producing metabolites that inhibit the growth of the pathogens and physically eliminating the pathogens from the plant by occupying the space and sites first. Microorganisms not naturally present in plant environment can be introduced in an attempt to control diseases (Cook, 1993).

This can be done by application of organic materials that contain natural microbial populations such as composts or natural microbial populations added to them including natural organic fertilizers with microbial supplements.

In both cases, the products must be applied prior to disease development as they are preventive and not curative (Baker, 1987; Cook, 1993). Natural organic fertilizers should be used for their nutritional value (nitrogen and potassium) and not for any possible secondary effects.

Fungal plant pathogens are very diverse and cause diseases on different parts of plants such as root, stem, leaf, fruit, etc. In this section, application of biological control strategies for controlling fungal diseases on different parts of plants will be discussed.

The majority of research on biocontrol of fungal diseases have focused on soil borne diseases rather than foliar or post harvest. According to the results of numerous research projects, several fungal and bacterial biocontrol agents have been used as seed and soil application to reduce the incidence of plant diseases caused by soil borne fungal pathogens (Cook, 1993; Heydari, 2007; Heydari *et al.*, 2004; Heydari and Misaghi, 2003; Lo *et al.*, 1995, 1996, 1997; McSpadden-Gardener, 2001; Naraghi *et al.*, 2004; Ramette *et al.*, 2003; Scheuerell *et al.*, 2005). Since many plant pathogens can spread readily in the foliar parts, control of these diseases requires both suppression of initial plant infection and reduction of the infection rate (Lo *et al.*, 1997).

Granular applications of strain 1295-22 of *Trihoderma harzianum* has been shown to significantly inhibit disease severity of some plant diseases during the initial stage of disease development, most likely by reducing levels of the pathogen inoculum in the soil (Lo *et al.*, 1995, 1996, 1997). It is apparent, therefore, that soil applications alone cannot effectively control the foliar phases of this disease.

Additives have been commonly used with fungicides to improve efficacy and they also may enhance the ability of biocontrol agents to reduce plant diseases.

For example, it was reported that seed treatment using 10% Pelgel with solid matrix priming markedly enhanced the efficacy of Trichoderma strains to control *Pythium* sp. on various crops (Lo *et al.*, 1997). Research has indicated that for control of multiple fungal plant diseases, greater control was obtained when Triton X-100 was included than when no additives, Pelgel, or Tween 20 were used (Lo *et al.*, 1997).

The use of specific surfactants with Trichoderma strains seems essential to obtain levels of control equivalent to those achieved with chemical fungicides. Detergents such as Triton X-100 may have several functions in biocontrol systems.

They may slow the growth of pathogens more than that of the biocontrol agents or they may enhance wetting and adhesion of spores to infection courts (Lo *et al.*, 1997).

In preliminary experiments, both Tween 20 and Triton X-100 slowed the growth of both *T. harzianum* and the pathogens, but the ratio of the growth rates of *T. harzianum* and pathogens was greater with Triton X-100 than with Tween 20 (Lo *et al.*, 1997).

Living organisms, in addition to yielding a large quantity of biomass of the bioprotectant fungus, must perform effectively in each application. To examine this, different spore formulations of *Trihoderma harzianum* were compared in a study for controlling plant diseases (Lo *et al.*, 1996).

It was found that all formulations provided equivalent levels of control, indicating that the method of spore production may not be a key factor in the efficacy of this fungal biocontrol agent in controlling these diseases (Lo *et al.*, 1996). To predictably and successfully use biological control agents for fungal disease control, it is critical that their biology and ecology be more completely understood. Therefore, effective antagonists must become established in plant ecosystems and remain active against target pathogens during periods favourable for plant infection.

Broadcast application of granules of Trichoderma to control plant diseases has resulted in establishment of stable and effective populations of plants in soils (Lo *et al.*, 1995, 1996, 1997). Similarly, it was shown that the populations of *T. harzianum* in soils treated with spray applications were as high as those in soils treated with granular formulations (Lo *et al.*, 1996). Population levels of strain 1295-22 in about 5x10 5 cfu g of soil significantly reduced Pythium blight, root rot and brown patch diseases (Lo *et al.*, 1997).

However, spray applications, even though resulted in numerically similar levels of root colonization, did not provide the same benefit. This may reflect the differences in inoculum potential of granules versus spray applications.

Granules are applied as a several-millimetre-diametre particle that is completely colonized by the fungus. Conidial inoculum, on the other hand, is much smaller and would therefore be expected to possess lower inoculum potential than the granular formulation (Lo *et al.*, 1997).

Conversely, in greenhouse and field experiments, it was found that *Trihoderma harzianum* significantly reduced some foliar phases of plant diseases when spray applications of conidial suspensions containing Triton X-100 were used (Lo *et al.*, 1995, 1996, 1997).

Weekly spray applications were as effective as the standard (monthly) fungicide applications. These results indicate that the efficacy of *T. harzianum* against plant diseases, especially those involving secondary infections, is very strongly affected by the method of application (Lo *et al.*, 1997).

The ability to survive on the plant phylloplane is also a desirable trait for strains of fungal and bacterial antagonists used as biocontrol agents against foliar diseases (Lo *et al.*, 1997). Spray applications of strain 1295-22 of *T. harzianum* has resulted in disease suppressive population levels on leaf (Lo *et al.*, 1997).

These populations were sufficient to suppress Pythium root rot, brown patch and dollar spot over the entire season. Thus, *T. harzianum* 1295-22 may possess a measure of phylloplane competence on the plants. The ideal biocontrol strategy attempts to introduce or promote the activity of biocontrol agents only when and where they are needed or are most effective and minimizes wasteful application of inoculum to non-target habitats (Lo *et al.*, 1997). Thus, for effective delivery, it is necessary to consider plant–pathogen–antagonist interactions in terms of time and space.

Pythium, Rhizoctonia and Sclerotinia are important soil borne pathogens of many plant species and their survival structures in soil serve as primary inoculum. Consequently, suppression of the initial inoculum will be the first step in managing these pathogens (Lo *et al.*, 1997).

The granular application of biocontrol agents should be followed by monthly spray applications to suppress foliar phases of these diseases. Inhibition of the secondary infection and dissemination of these pathogens is also important for disease management (Lo *et al.*, 1997).

Monthly spray applications of *T. harzianum* could provide a second step in protection of plant foliage from attack by preventing these pathogens from initially infecting leaves and by reducing the spread of disease or other methods of inoculum dissemination. Finally, results of Lo *et al.* (1997) study have indicated that it will be necessary to apply weekly sprays for highly effective control of these pathogens under severe disease conditions.

In addition to Trihoderma and other fungal antagonists, several antagonistic bacterial species including *Pseudomonas fluorescens*, *P. putida*, *P. aerofaciens*, *Burkholderia cepacia*, *Bacillus subtillis*, *B. Polymyxa* and *B. cerrues* have also been used successfully in biological control of different soil borne fungal diseases (Heydari *et al.*, 1997, 2004, 2007; Heydari and Misaghi, 2003; Kloepper *et al.*, 2004; Leeman *et al.*, 1995; Shahraki *et al.*, 2009; Shishido *et al.*, 2005; Weller and Cook, 1983; Zaki *et al.*, 1998). By application of these bacterial antagonists, various fungal pathogens including *Rhizoctonia solani*, *Fusarium moxysporium*, *F. solani*, *Verticillium dahliae*, *Gaummanno-muces graminis* and soil borne diseases caused by them such as seed rot, damping-off, root rot, vascular wilt and take-all have been biologically controlled on major agricultural crops including cotton, sugar beet, wheat, rice and different vegetables.

Although the majority of biological control research have been concentrated on soil borne fungal diseases, a number of studies have focused on fungal pathogens causing diseases and disorders in above-ground parts of plants (Kessel *et al.*, 2005; Khodakaramian *et al.*, 2008; Kovach *et al.*, 2000; Milgroom and Cortesi, 2004; Smith *et al.*, 1993).

For example, Anderson *et al.* (2004) studied the possibility of biological control of fungal pathogens in the phylosphere and proposed that it may be possible to reduce the incidence and development of these diseases using fungal and bacterial antagonists. In another study, biological control of powdery mildew disease on different crops using antagonistic fungi was investigated and it was found that biocontrol-active micro-organisms can potentially be applied against this very important foliar diseases. *Botrytis cinera* which is the causal agent of gray mold on many plants (Agrios, 1988) was successfully controlled by the use of biocontrol-active microorganism on strawberry (Kovach *et al.*, 2000). In another study conducted by Smith *et al.* (1993) biological control of cotton leak of cucumber caused by a fungal foliar pathogen was studied. It was found that *Bacillus cerrues,* a bacterial antagonist was capable of reducing the incidence of the disease significantly (Smith *et al.*, 1993).

Another example of using biocontrol-active microorganisms against foliar fungal pathogen is the study in which chestnut blight was successfully controlled by the virus that infects *Cryphonectria parasitica*, the fungal causal agent of the disease through the mechanism of hypovirulence, a reduction in pathogenicity of the pathogen.

This phenomenon has resulted in control of the chestnut blight in many places (Milgroom and Cortesi, 2004). However, the interactions of virus, fungus, tree and environment play very important role in the success of disease control.

In addition to soil borne and foliar diseases some studies have also tested the efficacy of biocontrol-active microorganisms on post harvest fungal pathogens which cause losses to fruits and vegetables during post harvest and storage periods (Janisiewicz and Korsten, 2002).

Spray applications of fungal and bacterial antagonists have resulted in significant reduction in the infection caused by some fungal pathogens in the storage.

The Use of Compost as Biofertilizer

Research data and observations in nurseries have shown that addition of composted organic matter to potting mixes results in suppression of soil borne fungal diseases (McKellar and Nelson, 2003; Paulitz and Belanger, 2001).

The concentration of suppressive microorganisms in compost amended substrates is very high, but greatly reduced in soils or potting mixes after the amendment (McKellar and Nelson, 2003; Paulitz and Belanger, 2001). As a result, predictive disease suppression models have been developed based on the composition and concentration of microbial biomass.

The effectiveness of composts in suppression of soil borne diseases is dependent on heat kill, organic matters decomposition, recolonization of compost by suppressive microorganisms following heat kill and physical and chemical factors (McKellar and Nelson, 2003).

Although previous works have focused on plant soil borne diseases, current research indicates that potting mixes containing composted organic materials which also have been inoculated with *Trichoderma hamatum* can be effective as a biocontrol alternative to foliar fungicides; however, the mechanism of this systemic type of induced resistance is not yet understood (McKellar and Nelson, 2003). Although the growers have traditionally relied on aged pine bark and composted biosolids to provide the potential for disease suppression, research indicates that composted animal manure have the potential to replace some of these components, but a consistent quantity and quality of these materials will need to be incorporated (McKellar and Nelson, 2003; Paulitz and Belanger, 2001). The maturity (stability) of the composted manure and its salinity largely determine its ability to induce suppression.

Commercialization of Biocontrol

Commercial use and application of biological disease control have been slow mainly due to their variable performances under different environmental conditions in the field (Fravel, 2005; Mercier and Lindow, 2001; Paulitz and Belengar, 2001; Wang *et al.*, 2003). Many biocontrol agents perform well in the laboratory and green house conditions but fail to do so in the field. This problem can only be solved by better understanding of the environmental parameters that affect biocontrol agents (Fravel, 2005; Mercier and Lindow, 2001; Paulitz and Belengar, 2001; Wang *et al.*, 2003).

In addition to this problem, there has also been relatively little investment in the development and production of commercial formulation of biocontrol-active microorganisms probably due to the cost of developing, testing, registering and marketing of these products (Heydari *et al.*, 2007; Ardakani *et al.*, 2009).

Biological control agents are generally formulated as wetable powders, dusts, granules and aqueous or oil-based liquid products using different mineral and organic carriers (Ardakani *et al.*, 2009).

Currently in the market, a number of biologically based products are being sold for the control of fungal plant diseases (Ardakani *et al.*, 2009). A growing number of companies are also developing new products that are in the process of registration. Many of these companies are small, privately owned firms with a limited product-line.

Others are publicly traded and have substantial capitalization values. In addition, larger companies with more diverse product lines that include a variety of agrochemicals and biotechnological products have played a significant role in the development and marketing of products for the control of plant pathogens (Ardakani *et al.*, 2009).

Biocontrol products are either marketed as stand-alone products or formulated as mixtures with other microbials. Some products with biocontrol properties may not be registered, but are sold instead as plant strengtheners or growth promoters without any specific claims regarding disease control (Ardakani *et al.*, 2009). To help improve the global market perception of biope-sticides as effective products, the biopesticide Industry Alliance is establishing a certification process to ensure industry standards for efficacy, quality and consistency.

To improve commercial use and application of biological disease control it is extremely important to emphasize and concentrate on several factors including training of growers, formulation of biocontrol microorganisms and studying the role of environmental factors.

Future Outlook

Biological control really developed as an academic discipline during the 1970s and is now a mature science supported in both the public and private sector. Research related to biological control is published in many different scientific journals, particularly those related to plant pathology and entomology.

Additionally, there are some academic journals specifically devoted to this disipline. In the United States, research funds for the biological control are provided primarily by several USDA programs. These include the Section 406 programs, regional IPM grants, Integrated Organic Program, IR-4 and several programs funded as part of the National Research Initiative (Bloom *et al.*, 2003). Monies also exist to stimulate the development of commercial ventures through the small business innovation research programs. Such ventures are intended to be conduits for academic research that can be used to develop new companies (Spadaro and Gullino, 2005).

Much has been learned from the biological control research conducted over the past forty years.

But, in addition to learning the lessons of the past, biocontrol researchers need to look forward to define new and different questions, the answers to which will help facilitate new biocontrol technologies and applications.

Currently, fundamental advances in computing, molecular biology, analytical chemistry and statistics have led to new research aimed at characterizing the structure and functions of biocontrol agents, pathogens and host plants at the molecular, cellular and ecological levels (Spadaro and Gullino, 2005). Some of the research criteria that will advance our understanding of biological control and the conditions under which it can be most fruitfully applied are as follows:

Ecology of Antagonistic Microbes: Ecological factors play very important roles in the performance and activity of biocontrol-active microorganisms. In this regard, the following criteria need to be clarified and studied:

- The distribution of fungal pathogens and their antagonists in the environment
- The optimum conditions in which biocontrol micro-organisms exert their suppressive capacities
- The response of native and introduced populations to different management practices
- The determinants factor of successful colonization and expression of biocontrol traits
- The components and dynamics of plant host defense induction

Application Methods: In regard with application strategies still there are some areas which should be investigated and developed for the enhancement of the effectiveness of biocontrol microorganisms. These areas are as follows:

- The search for more effective strains or strain variants for current applications
- The use of genetic engineering of microbes and plants for enhancing biocontrol application methods
- The development of proper formulations to enhance activities of known biocontrol agents

Introducing new strains and mechanisms: Since fungal plant pathogens are very diverse and their pathogenicity is different on host plants, it is therefore very important to look for new and novel biocontrol microorganisms with different mechanisms. In this regard, the following criteria need further investigation:

- The use of previously uncharacterized microbes as biological control agents
- Study on the roles of other genes and gene products which are involved in pathogen suppression
- The efficacy of using novel strain combinations in comparison with individual agents
- Study on the signal molecules of plant and microbial origin which regulate the expression of biocontrol traits by different agents

Integrated Pest Management: Since the ultimate goal of biological control of plant diseases is to assist the growers to combat and control plant pathogens in the field which is the real agricultural environment, it is therefore important to practically integrate biocontrol strategies into agricultural system. In this regard, the following criteria should be considered and followed carefully:

- Selection of production systems that can most benefit from biocontrol for disease management
- Application and use of biocontrol strategies which best fit with other IPM system components
- Development of effective biocontrol-cultivar combinations by plant breeders

Research and development: Nowadays, growers are interested in reducing dependence on chemical inputs, so biological controls (defined in the narrow sense) can be expected to play an important role in Integrated Pest Management (IPM) systems (Jacobsen *et al.*, 2004). Good agricultural practices (GAP) including appropriate site selection, crop rotations, tillage, fertility and water management, provide the foundation for successful pest management by providing a fertile growing environment for the crop.

The use of disease-resistant varieties, developed through conventional breeding or genetic engineering, provides the next line of defense. However, such measures are not always sufficient to be productive or economically sustainable.

In such cases, the next step would be to deploy biorational controls of diseases. These include BCAs, introduced as inoculants or amendments, as well as active ingredients directly derived from natural origins and having a low impact on the environment and non-target microorganisms (Guetsky *et al.*, 2001; Jacobsen *et al.*, 2004). If these foundational options are not sufficient to ensure plant health and/or economically sustainable production, then less specific and less harmful synthetic chemical toxins can be used to ensure productivity and profitability.

With the growing interest in reducing chemical inputs, companies involved in the manufacturing and marketing of BCAs should experience continued growth. However, stringent quality control measures must be adopted so that farmers get quality products. New, more effective and stable formulations also will need to be developed.

Most fungal pathogens are susceptible to one or more biocontrol strategies, but practical implementation on a commercial scale has been constrained by a number of factors. Cost, convenience, efficacy and reliability of biological controls are important considerations, but only in relation to the alternative disease control strategies.

Cultural practices (e.g., good sanitation, soil preparation and water management) and host resistance can go a long way towards controlling many diseases, so biocontrol should be applied only when such agronomic practices are insufficient for effective disease control.

As long as petroleum is cheap and abundant, the cost and convenience of chemical pesticides will be difficult to surpass. However, if the infection court or target pathogen can be effectively colonized using inoculation, the ability of the living organism to reproduce could greatly reduce application costs. In general, although, regulatory and cultural concerns about the health and safety of specific classes of pesticides are the primary economic drivers promoting the adoption of biological control strategies in urban and rural landscapes (Timms-Wilson *et al.*, 2004).

Self-perpetuating biological controls (e.g., hypovirulence of the chestnut blight pathogen) are also needed for control of diseases in forested and rangeland ecosystems where high application rates over larger land areas are not economically feasible. In terms of efficacy and reliability, the greatest successes in biological control have been achieved in situations where environmental conditions are most controlled or predictable and where biocontrol agents can preemptively colonize the infection court (Fravel, 2005). Monocyclic, soilborne and post-harvest fungal diseases have been controlled effectively by biological control agents that act as bioprotectants (i.e., preventing infections).

Specific applications for the high value crops targeting specific diseases (e.g., downy mildew, powdery mildew and several other fungal diseases) have also been adopted (Kessel *et al.*, 2005). As research unravels the various conditions needed for successful biocontrol of different fungal diseases, the adoption of BCAs in IPM systems is bound to increase in the years ahead.

Conclusions

Due to the serious environmental and health problems that wide spread use of chemical pesticides has created in the world, search for alternative safe methods is unavoidable. Biological control of plant diseases has been the subject of numerous research projects in recent years (Bargabus *et al.*, 2004; Benhamou, 2004; Chisholm *et al.*, 2006; Heydari, 2007; Islam *et al.*, 2005).

There is a growing demand for biologically based pest management practices. Recent surveys of both conventional and organic growers indicate an interest in using biocontrol products suggesting that the market potential of biocontrol products will increase in the future (Joshi and Gardener, 2006). Application of different biological control strategies has been successful in the greenhouse industry and continues to increase (Jacobsen *et al.*, 2004).

An upswing in commercial interests has also developed in the past few years and prospects for increased growth are positive. The Biopesticide Industry Alliance has formed and it is now actively promoting the value and efficacy of biopesticides (including those that control fungal plant pathogens). Clearly, the future success of the biological control industry will depend on innovative business management, product marketing, extension education and research (Timms-Wilson *et al.*, 2004; Joshi and Gardener, 2006). Increased demand for organic products in home gardening activities by using non-chemical methods has enlarged the market for biocontrol products. The field of plant pathology will contribute substantially to making the 21st century the ages of biotechnology by the development of innovative biocontrol strategies. A variety of research questions remain to be fully answered about the nature of biological control and the means to most effectively manage it under production conditions. Advanced molecular techniques are now being used to characterize the diversity, abundance and activities of microbes that live in and around plants, including those that significantly impact plant health (Joshi and Gardener, 2006). Still, much remains to be learned about the microbial ecology of both plant pathogens and their microbial antagonists in different agricultural systems. Fundamental work remains to be done on characterizing the different mechanisms by which organic amendments reduce plant disease including those caused by fungal pathogens.

More studies on the practical aspects of mass production and formulation need to be undertaken to make new biocontrol products stable, effective, safer and more cost-effective.

Fungal pathogens are among the most important factors that cause serious damages and losses to plants. Harmful impacts of the chemical pesticides on the environment and non-target organisms have clearly been documented. The need for the development of non-chemical alternative strategies to protect plants against plant diseases including fungal pathogens is therefore clear. Biological control using fungal and bacterial antagonists to manage plant diseases seems to be a promising alternative strategy and have successfully been applied to control some diseases on different plants and crops. Biocontrol strategies may also be used to manage other plant diseases including foliar ones. Some of the important factors that affect the efficacy of microbial biocontrol agents in controlling plant diseases which should carefully be considered include method of application, formulation of biocontrol microorganisms and timing of application. Various composts and organic amendments as other means of biological control have also been tested on some plants and proven to be promising.

There are many products composed of living organisms, primarily bacteria and fungi, being sold that claim they will increase plant health. However, for any material to be considered a biological fungicide the Environmental Protection Agencies and Organizations must register it (Bloom *et al.*, 2003). This registration indicates that the safety of the product to humans, non-humans (fish for example) and the environment has been determined. Materials that have not been approved should be used with caution. Complete elimination of chemical pesticides for controlling plant pests and diseases in modern agriculture may be impossible, but a logical reduction in their application is absolutely feasible. To have a sustainable agricultural system with minimum contamination and risks to the environment, a combination of all available methods should be applied to manage pest problems and this can be achieved by Integrated Pest Management (IPM). Implementation of IPM strategies may be the safest solution for management of pest problems including fungal diseases in every cropping system and with no doubt biological control is one of the most important components of Integrated Pest Management which can lead us toward a sustainable agricultural system in the future.

Crop Disease and Management

History and Importance of Plant Bacteriology

History: History of bacteriology started when Anton Von Leeuwenhoek, (Dutch Worker) observed Bacteria for the first time in 1675 with Microscope. In 1683 he described the bacteria seen with microscope. The importance of these observations were realised only after Pasteur (1876) had demonstrated their role in fermentation and decay.

Robert Koon (1876) from Germany had proved that the bacteria can cause such disease as anthrax, tuber culosis or Asiatic cholera. He gave four famous Koch's Postulates for proving that a particular organism is the cause of particular disease.

Woronin (1876) isolated and described the root nodule bacteria in leguminous plant. Now it was well recognised that all the organisms' bacteria are most closely related to human being life.

T.J Burbil from University of Illinois (1878) was first proved to association of bacterium with plant disease. He showed that Erwinia amylovora causes fire blight pear and apple.

In 1879 Prilleaux reported the bacterial decay of wheat kernels.

In 1883 J.H Wakker published the results on his through investigationson yellow slime disease of hyacinth caused by bacterium.

J.C Arther (From USA) during 18851887 confirmed Burrill's work.

Luigi Savastano (From Italy) described in 1887 a bacterial knot disease of olives.

Erwin F. Smith (From USA) started studying bacterial disease of pants from 1890. His humerous and excellent contributions on to the study of bacterial disease of plants, particularly of the bacterial wilts of cucurbits, of solanaceous crops, and of cruifers established beyond any doubt the role of bacteria as phytopathogens.

He resolved the controversy with the German Bacteriologist, Alfred Fischer (1897,1899) who did not think that bacteria were the primary cause of disease in plants. He also was among the first to notice (1893-1894) and to study the crown gall disease. He showed it to be caused by bacteria, studied its anatomy and development, and considered it to resemble cancerous tumors of human and animals.

Chilton et al. (1977) showed that the crown gall bacterium transforms normal plant cells into tumor cells by introducing into them a plasmid, part of which becomes inserted into the plant cell chromosomes DNA.

Windsor and Black in 1972 observed rickettsia like organisms in the phloem of clover plants infected with the club leaf disease. The following year similar organisms were observed in grape infected with pierces disease, in peach infected with phony peach, and others.

E.J Butler from India, examined the brown rot disease of potato in 1903, and indicated the bacterial nature of the disease.

M.K Patel and associates from Pune, India, commensed a serious of studies on bacterial phytopathogens in 1948. This team, over a period of about 15 years, reported on nearly 40 bacterial diseases, some of them on economically important host plants.

Importance: About 1600 bacterial species are known. Most are strictly saprophytes which are beneficial to human because they decompose the enormous quantities of organic matter produced yearly by humans, animals and factory waste products or by death of plants and animals and factory waste products or by death of plant and animals. Several species causes disease in humans, animals and plants.

General Characters of Plant Pathogenic Bacteria

1. Almost all plant pathogenic bacteria are rod shaped, the only exception being Streptomyces, which is filamentous.
2. The rod shaped bacteria are more or less short and cylindrical and in young cultures, they range from 0.6 to 3.5 m in length and from 0.5 to 1.0 m in diametre.
3. In older cultures or at high temperatures, the rods of some species are much longer and they may even appear filamentous.
4. Sometimes deviations from the rod shape in the form of a club, a Y or V shaped, and other branched forms occur, and some bacteria may occasionally occur in pairs or in short chains.
5. The cell walls of bacteria of most species are enveloped by a viscous, gummy material, which may be thin (Slime layer) or may be thick, forming a relatively large mass around the cell (Capsule).

6. Most plant pathogenic bacteria are equipped with delicate, thread like flagella.
7. In some bacterial species each bacterium has only the flagellum, others, have a tuff of flagella at one end of the cell (Polar flagella); some have a single flagellum or a tuft of flagella at each end, and till others have peritrichous flagella, that is, distributed over the entire surface of the cell.
8. In the filamentous Streptomyces species, the cells consist of non septate branched threads, which usually have a spiral formation and produce conidia in chains on aerial hyphae.
9. Single bacterium appears hyaline or yellowish white under the compound microscope.
10. Bacteria grow and produce colonies on solid medium.
11. Colonies of different species may vary in size, shape, form of edges, elevation and colour, and are sometimes characteristics of a given species.
12. Bacterial cells have thin, relatively tough, and somewhat rigid cell walls.
13. All the material inside the cell wall constitutes the protoplast.
14. The nuclear material consist of a large circular chromosome composed DNA and appear as spherical, ellipsoidal or dumbbell shaped body within the cytoplasm.
15. Often bacteria also have single or multiple copies of addition smaller circular chromosomes called 'Plasmids' that can move or be moved between bacteria or between bacteria and plants as for example in the crown gall disease.
16. Rod shaped Phytopathogenic bacteria reproduce by the asexual process known as binary fission or fission. Under favourable conditions bacteria may divide every 20 minutes.
17. Almost all plant pathogenic bacteria develop mostly in the host plant as parasites and partly in plant debris or in the soil as saprophytes.

Classification of Phytopathogenic Bacteria

According to the 8 th Ediction of Bergy's Manual Phytopathogenic bacteria are classified as:

Division II: Sctobacteria (indifferent to light)

Class I: The Bacteria

Part: Gram negative aerobic rods and cocci.

Family: Pseudomonaceae

Genera: 1. Pseudomonas

Pseudomonas

Characters: Cells single, straight or curved rods but not helicie, generally 0.5 X 1.5 to 4 microns in size, motile by polar flagella one or more in number, non spore forming, gram negative; strict aerobes; chemoorganotrophs; metabolism respiratory, never fermentative.

Species

1. Pseudomonas solanacearum causes brown rot or wilt of potatoes.
2. Pseudomonas rubrillineans causes red stripe of sugarcane.

Xanthomonas

Characters: Cells single, straight rods measuring 0.2 to 0.8 X 0.6 to 2.0 microns (Usually 0.4X 1.0 microns); Gram – ve, motile by a single polar flagellum; non spore forming; copious extracellular slime produced; strict aerobes; metabolism respiratory, never fermentative ; oxidase negative and catalase positive; bacterial colonies yellow coloured due to production of Xanthomonadin pigment.

Species

1. X. campestris pv. Citri causes citri cause citrus canker.
2. X. campestris pv. Malvacearum causes angular leaf sport and black arm of cotton.
3. X. campestris pv. Oryzae causes bacterial leaf blight of rice.

Family: Rhizobiaceae

Genus: Agrobacterium

Character: The bacteria are rod shaped, 0.8 X 15. m. They are motile by means of 1.4 peritrichous flagella, when only one flagellum is present it is more often lateral than polar.

When growing on carbohydrate – containing media the bacteria produce abundant polysaccharide slime. The colonies are non-pigmented and usually smooth. These bacteria are rhizosphere and soil inhabitants.

Species

Agrobacterium tumefaction causes crown gall of pome and stone fruit trees, brambles and grapes. A. rubi causes cane gall of raspberries and black berries. A. rhizogenes causes hairy root of apple.

Gram Negative, Facultative Anaerobic Rods

Family: Enterobacteriaceae

Genus: Erwinia

Characters

Cells predominantly single, straight rods, measuring 0.5 to 1.0 X 1.0 to 3.0 microns; motile (Except E.stewartii and E.dissolves) by peritrichous flagella; gram negative ; grows well on artificial media aerobically as well as anaerobically and produce acid.

The genus has three major groups of species.

- Pectolytic bacteria that cause oft rot.
- Those do not cause soft rot but dry necrosis and wilt and
- Epiphytes that cause neither soft rot nor necrosis.

Species

1. E. amylovora causes fire blight of apple.
2. E. tracheiphia causes vascular wilt of cucurbits.
3. E. carotovora pv. atroseptica causes black leg of potato.
4. E. carotovora pv. Carotovora causes soft rot numerous fleshy fruits, vegetables and ornamentals.

Chapter 5

Symptom Caused by Phytopathogenic Bacteria

Local Lesions

Localised spots may occur on the leaf blade, petiole, stem and fruits. The leaf spots starts as minute water soaked specks, spreading rapidly to form circular, irregular or angular spots, bound by veins and veinlets.

The pathogen which enters the host tissues through stomatal openings establishes in the substomatal space and remains so in the parenchymatous tissues causing necrotic lesions. E. g Angular leaf sop of cotton caused by Xanthomonas campestris pv. malvacearum.

Blights

The invasion by the bacterium loads to very rapid and extensive necrosis of the affected plant parts resulting in scorched appearance of the host.

E. g Bacterial blight of paddy caused by Xanthomonas campestris pv. oryzae.

Soft Rots

The soft rot symptoms are found mainly on fleshy parts. The major effect is softening of the tissues due to disintegration of cells and dissolution of middle lamella, as a result of enzyme action.

Very often a dirty liquid oozes out of the affected parts. In many causes, the disintegration is preceded by change of colour. E.g Black leg and soft rot of potato caused by Erwinia carotovora sub species carotovora and Erwinia carotovora sub species atroseption.

Tumors and Galls

In many bacterial diseases the effect of invasion by the pathogen is hyperplasia and hypertrophy of invaded tissues. As a result, tumours develop on the affected organs. E.g Crown gall tumors on rose plant caused by Agrobacterium tumefaciens.

Vascular Disease

In same of the bacteria leaf spot diseases the organism moves into the vascular system and becomes systemic. In others, the invasion is concentrated in the vascular tissues causing typical wilt of the plant by plugging the water conducting vessels and by production of toxins.

E.g Bacterial canker and wilt of tomato caused by Clavibactor (Corynebacterium) michiganese sub species michiganese. Bacterial wilt of tomato, potato, tobacco and egg plant caused by Pseudomonas solanacearum.

Scabs and Cankers

Scabs and cankers are corky outgrowth which is formed on leaves, twigs and all other plant parts above the ground. These outgrowths are the result of the reactions of the host tissues to the pathogen.

Such reactions are mostly localised and often confirmed to the parenchymatous tissues of the host plant.

Scab is formed by epidermal infection and is not deep seated. Cankers on the other hand, are deep seated and involve the cambium layer.

E. g 1. Citrus canker caused by Xanthomonas Campestris pv. citri. Common scab of potato caused by Streptomyces scabs.

Mode of Entry of Bacteria in Plants

Unlike fungal pathogens, bacteria are incapable of mechanically penetrating the cultinized plant tissues, culticle, periderm, etc. Since the ability to form appresoria is lacking in them. The ways of gaining entry into the plants are as follows.

1. Active invasion by bacteria through non-cultinized surface is sometimes feasible by means of a large mass of cells functioning together. Rhizobium spp. Which causes nodules in legume form a muciliagenous bacterial colony or zoogloea at the apex of the root hairs or root tips, the Zoogloea dissolves the cell wall and then, in the form of a curved and branched infection filament or infection thread advances into the interior of root hair and hence, into the root corner.
2. The cells of black rot of caggage, Xanthomonas campestris, enter through hydathodes, which are specialized gland cells at leaf edge.
3. The fire blight bacterium, Erwinia amylovora, enters the floral parts through the special nectar producing cells present system.
4. The potato scab organism, Streptomyces scabies, gains entry into the tubers through lenticels.
5. Entry through stomata: The mechanism of entry into the stomato is through water congestion which leads the cells into the open stomato assisted by wind blown rain or dew.

E.g

1. Pseudomonas tabaci causing wild fire of tobacco.
2. Xanthomonas. campestris pv. malvacearum causing angular leaf spot of cotton.
3. Xanthomonas. Phaseoli causing bean blight.
4. Xanthomonas. Campestris causing black rot of cabbage.

6. Entry through Wounds:

During various cultivation operations and while harvest, transport and storage, injuries to various plants parts are caused. Many bacteria present in vicinity enter the host tissue.

E .g Erwinia spp. causing soft rot of vegetables.

Agrobacterium tumefaciens causing crown gall of fruit trees.

7. Entry through Insect Wounds:

E. g Crown gall of fruit trees and wilt of cucumber.

8. Entry through wounds caused by plant parasitic nematodes and funi. Root disease complex in citrus.

Important Phytopathogenic Bacteria

Sr.No	Genus	Family	Gram Reaction	Characters	Symptoms produced on host	Example
1	Pseudomonas	Pseudomona deveae	-Ve	1. Rod straight to curved, size 0.5-1 X 1.5-4 mue m, one to many polar flagella.2.Colonies not yellow, do not produce acid from lactose.	Leaf spot, blights, vascular wilts, soft rots, canker.	1. Brown rot or bacterial wilt of potato.2 Bacterial wilt of brinjal and tomato.
2	Xanthomonas	Pseudomona deveae	-Ve	1. Straight rods, size 0.4-1X 1.2 -3m. One polar flagellum present.2. Colonies yellow due to Xanthomonadin , produce acid from lactose.	Leafspot, fruit spots, blight, canker	1. Citrus canker, 2. Black arm of cotton,3.Blight of paddy.
3	Agrobacterium	Rhizobiaceae	-Ve	1. Rod shaped, size 0.8 X 1.5 -3 mue m. 1 to 4 peritrichous flagella.2. Colonies mostly white do not hydrolyze starch.	Crown Gall	Crown Gall of stone fruit roses.
4	Erwinia	Enterobacteriaceae	-Ve	1. Straight rods, size 0.5-1, 0-3.0 mue m , several peritrichous flagella.	Fire blight, wilt soft rot.	1. Fire blight of apple.2. Soft rot of vegetables.
5	Clavibactor (Corynebacterium)	Enterobacteriaceae	+Ve	Straight to straightly curved rods, 0.5-0.9X 1.5 - 4.0Mue m. non motile but some species are motile by one or two polar flagella.	Wilt	Wilt of potato ad Tomato
6	Streptomyces	Streptomycetaceae	+Ve	Slender branched hyphae without	Scab	Potato scab

Perpetuation of Bacteria

The Phytopathogenic bacteria do not produce any resting structures. They survive by some of the following ways:

Survival on Self Sown Plant, Collateral Hosts: Bacterial pathogen survives and multiplies on self sown plants of host crop, collateral hosts during the main and the off seasons.

The collateral hosts are most often perennial or noxious weeds growing in the host crop fields, or in their vicinity while others may be cultivated economic crop plants. The host range of some bacterial plant pathogens is very wide.

E. g Pseudomonas solaneceanum -200 host plants.

Agrobacterium tumefaciens- 150 host plants.

Survival on Host Crop: The pathogens of the bacterial leaf blight and the leaf streak of rice may survive from crop to crop in double or triple – rice culture areas in the tropics. The latent infections as in cankers, galls, bud scales, the cracks in the bark or sheltered place are often further protected by the host mucilage in trees as in the citrus canker or the fire blight.

Survival on Non Host Plants: Xanthomonas campestris pv. Citri causal agent of citrus canker has been found to survive in the non-host plants. The epiphytic (Saprophytic) existence of X.campestris pv . phasebli on the phyllosphere of Phaseolus vulgaris has been found.

Survival through Crop Residues and in Soil: Some of the Phytopathogenic bacteria are capable of living saprophytically on plant residues or parennate in the soil in Free State or on any organic matter. E.g Erwinia spp. Pseudomonas, solanacearum racee2, Xanthomoans campestris pv. malvacearum. Most of the bacterial pathogens of aerial parts can not survive in a natural soil for a long time due to their poor competitive saprophytic ability and antagonism of other microorganisms.

Survival through Seeds and Vegetative Propagation Parts: Some of the phyopathogenic bacteria survive in seeds of some crops and vegetative plant parts used for propagation E. g Xanthomonas campestris pv. oyzae and Xanthomonas compestris pv. malvacearum survive through seeds. Pseudomonas solanacearum survive in potato tubers.

Transmission (Dissemination) of Bacteria

The bacterial diseases are mainly disseminated through the agency of seed, air and water.

Seed: Xanthomonas campestris pv. oryzae (in Paddy), X.campestris pv. malvacearum (in cotton) are carried either externally or internally through seed. Corynebacterium sepedonicum and pseudomonas solanacearum (in sugarcane) are spread through vegetative plant parts used for propagation.

Soil: Phytopathogenic bacteria live as saprophytes on plant residues in the soil and when suitable host plants are sown they become active pathogens.

They are disseminated from place to place by means of cultivation operations, tool of cultivation and related agencies-human and animal and also through irrigation water, rain water and wind E.g of disease spread: Bacterial blight of cotton (X. campestris pv. malvacearum).

Wind: Bacteria surviving on fallen leaves etc and in soil are carried in the wind for long distance and deposited on the host plants.

Rain and Water: Rain drops falling on bacterial pustules or oozo present on the host surface distribute the bacteria in fine droplets to to other portions of the plant. The motile ones can move short distances in drops of water on plant parts and gain entry into the plant tissues.

Rain water especially when accompanied by wind, forms an important agent for rapid spread of the pathogens.

Bacteria in soil are carried via rain or irrigation water. E.g of disease spread: leaf blight of rice Xanthomonas campestris pv. oryzae.

Insects: Insects play a major role in the dissemination of bacteria. They act as carries of bacterial cells as well as help in infection through wounds E.g disease spread through insects: Fire blight of apple (Erwinia amylovora).

Mechanical: Use of contaminated tools and bags during cultural operations and movement of infective material by implements and other agencies during various field operations also play an important role in the spread of bacterial disease e.g of disease spread: Brown rot of potato (Pseudomonas solanacearum)

Man: Man is the agent behind the long distance dissemination which takes place chiefly through infected seeds and vegetative propagating material. E. g of disease spread: Leaf blight of paddy (Xanthomonas. compestris pv. oryzae), brown rot of potato (Psudomonas solanacearum).

Management of Bacterial Diseases

Bacterial plant pathogens directly enter the host tissue either through natural openings or through wounds; therefore, it is difficult to manage them. Some of the general principles of managing the bacterial disease are as follows:

Exclusion: Exclusion of a disease entry either from a foreign nation or within the country, fro a diseased to a healthy tract is an effective measure for avoiding the disease. Quarantine measures are in vogue in several countries. E .g diseased managed: Citrus canker (Xanthomonas campestris pv. citri) in U.S.A.

Eradication

Eradication is generally carried out for eliminating a well established pathogen and its host plant, collateral hosts, or insect vectors and some times use of cultural measures for starving out or killing the pathogen E. g of disease managed.

- The complete distruction of all infected trees was successfully achieved the citurs canker in Florida, USA during 1914-1927.
- Pruning of the infected twigs is followed to reduce the inoculums in orchards for citrus canker.
- The destruction of volunteer plants and weed hosts brings down the inoculum level. e.g Brown rot of potato (Pseudomonas Solanacearum).
- The crop residue may be burnt or ploughed deep into the soil with watering to ensure decomposition, which is helpful for the pathogens which cannot live saprophytically in the soil. E. g Bacterial blight of cotton (Xanthomonas campestris pv. malvacearum).
- Crop rotation with cereals is advocated for the management of wilt of tobacco (Pseudomonas solanacearum).
- The selected application of NPK , particularly low and split doses of N, helps to reduce the intensity of the bacterial leaf blight of rice (Xanthomonas campestrsi pv. Oryzae).
- The control of insects is helpful in reducing the soft rot of vegetables (Erwinia spp.) and the citrus canker (Xanthomonas campestris pv. citri).

- The avoidance of cultural mismanagement favourable to disease is important such as water logging in the nurseries (against the bacterial leaf blight of rice), flooding or over irrigation in the field. (against the soft rot disease)
- The sterilization of the cutting knife by flame or by 0.1 % KMNO4 solution while cutting the potato tuber for sowing is recommended against the potato wilt.
- The seed certification programes to raise pathogen free seeds. E.g bean (Xanthmonas campestris pv. campestris) are in practise in U.S.A.
- Elimaknation of the extremally and internally seed born pathogen by seed treatment. E. g i) Delinting of cotton seed with concentrated sulphuric acid for bacterial blight of cotton. ii) Hot water treatment of cotton seed at 560C for 10 minutes for bacterial blight of cotton. iii). Soaking of rice seeds in 0.025% strpetocyclic solution for leaf streak disease of rice.
- When tobacco is immediately grown after maize there is a considerable reduction in the incidence of Pseudomonas solanacearum on the later host.
- There is some prospect of biological control with the application of organic matter in the form of compost and green manure and even inoculation of antibiotic micropropaganism, encouraging antagonism by the application of superphosphate, application of phages in seed plant and soil as well as by inoculating with bacteriocin- producing strains.

Plant Sanitation

These measures are adopted so as to avoid the disease onset and also to prevent spread.

E.g. i) Removal of dead plant parts from fruit crops and protection of cut surface with suitable bactericides (E. g for citrus canker and fire blight of apple). ii) Avoiding injuries to the plant parts at the time of cultural operations and during harvest, transport and storage. iii) Collection and destruction of diseased fallen leaves, bolls etc. for bacterial blight of cotton.

Plant Protection

These measures are adopted to prevent the onset and subsequent spread of plant diseases. E.g

(i) Seed treatment of cotton and rice with antibacterial chemicals against bacterial blight and leaf streak diseases respectively.

(ii) Treatment of vegetatively propagation material with bactericides e.g 0.02% stroptocycline for 30 minutes treatment of seed tubers of potato against brown rot disease.

(iii) Foliar sprays with Bordeaux mixture and copperoxy – chloride against leaf spots and blights.

(iv) Foliar sprays of streptomycin sulphate, 100 and 500 ppm against fire blight of apple and citrus canker.

(v) *Therapy:* Pancillin and vancomycin are effective in the disintegration of the crow gall. The application of antibiotics for protection also involves therapy.

The bacteriocines have also been demonstrated to be successful against bacterial disease E. g Agrobacterium radiobacter against crown gall of peach and tomato seedlings (Agrobacterium fumetaciens.)

(vi) *Immunisation:* By screening under artificial epiphytic conditions, resistance source for bacterial pathogens can be known. Resistant varieties are evolved by selection, breeding and other methods. E. g HC-9, BJA-592, P-14, T-12, 101-102 B, Reba-B-50, Khandwa-2, DHY-286, B-1007, cultivars of cotton and N-22, IR-22 cultivars of rice resistant to bacterial blight disease of these crops.

(vii) *Disease Forecasting:* Plant disease forecasting help in timely and economic management of bacterial diseases. Following basic principles are used for forecasting.

(a) *Forecasting by natural infection:* This is based on observation of plants at regular intervals- both in the nursery as well as in the main field for the occurrence of the disease.

(b) *Forecasting by climatic condition:* Effect of various metrological parameters on the incidence and spread of the disease in studied in detail and the congeneal meteorological parameters for these are workedout. Prediction model is prepared on the basis of these observations.

(c) *Forecasting by bacteriophage population:* A correlation has been obtained between the occurrence of bacterio-phage in the field (Soil, irrigation water, etc.) and disease outbreak in the case of bacterial leaf blight of rice.

Prior to outbreak of the disease the phage population in soil and irrigation water increase. It is possible, to some extent to forecast the disease outbreak using the phage population as an index.

Definition and Characteristics of Virus

Definition of Virus: Mathwas (1981) considers a virus as a set of one or more template molecules normally encased in a protective coat or coats of protein or lipoprotein, which is able to organize its own replication only within suitable host cells where its production is:

- Dependent on hosts protein synthesizing machinery (ribosomes).
- Organised from pools of required material rather than binary fission and
- Located at sites which are not separated from the host cell contents by a lipoprotein bilayer membrane.

Bos (1983) defines virus as an infectious agent often causing disease, invisible with the light microscope (Sub-microscopic) , small enough to pass through a bacterial filter, lacking a metabolism of its own and depending on a living host cell for multiplication.

Viruses are small packages of host alien genetic information of one type. (RNA or DNA), either in one strand or in a few segment. Encapsulated together or separately and enclosed in a coat of one or more types of protein, some time with an extra coat (envelope) and some other constituents.

Characteristic of Virus

Following characters will always be includes:

- Viruses contain one or more pieces of a single type of nucleic acid, either RNA or DNA, never both.
- The nucleic acid carries the genome of the virus which differs from one virus to another.
- The genome in the nucleic acid strand directs the synthesis of specific proteins for the protein coat which must be present in all viruses throughout their active phase except at the time of replication when, protein coat and nucleic acid are separated.
- Viruses rely on living host cells for most of the enzymes necessary for their replication.
- Viruses are in cable of growing for the synthesis of Lipman system is absent in Viruses.

History of Plant Virology

1. Virus disease of plants was known long before the discovery of bacteria.
   (i) Breaking of flower colour of tulips (as early as 1576).
   (ii) Transmission of leaf variegations from the scion to the stock of woody plants (as early as 1700).
2. Tobacco mosaic was identified by Swietch in Holland in 1857.
3. Adolph Mayer (1886) was the first to point out that tobacco mosaic is readily transmissible and infectious. Believed that bacteria are the cause of the disease.
4. Iwanowski (1892) confirmed some of the results Mayer. He demonstrated that the power to infect was lost if the sap was previously heated. He reported that infectiousness was retained even when sap was passed through bacteria proof filters.
5. Bejierink (1896) demonstrated that tobacco mosaic infectious agent could diffuse through an anger membrane and concluded that tobacco mosaic was caused by a non-corpuscular ‘ Contagium vivum fluidum’ which he called as Virus.

6. Hashimoto (1894) showed transmissible lity of rice dwarf disease by leaf hopper (Nephotettix apicalis var. cincticeps).
7. Stanley (1935) crystallized tobacco mosaic virus with ammonium sulphate and concluded that the virus was an autocatalytic protein that could multiply within the living cells.

   He considered that virus was a globulin containing no Phosphorus. For his discovery he was awarded a Nobel Prize.
8. Bawden and Pirrie (1936) described the isolation from TMV infected plants of a liquid crystalline nucleoprotein containing nucleic acid of the pentose type. They showed that the particles were rod shaped.
9. Kausche, Pfankuch and Ruska (1939) saw virus particles for the first time with the electron microscope. They confirmed that TMV was rod shaped.
10. Muller (1942): Williams and Wycoff (1944) developed shadow casting technique with heavy metals which was useful for determining the overall size and shape of the virus particles.
11. Markham and Smith (1949) isolated TYMV and showed that it contained (1) An infectious nucleoprotein (about 25%) and (2) non inferious protein.
12. Morel and Martin (1952) showed that virus free plants could be obtained from totally infected parents using meristem tip culture.
13. Kassanis (1954) showed that virus could be eradicated from infected plants by high temperature treatment.
14. Gierer and Schramm (1956) showed that the protein could be removed the virus and that the nucleic acid carried the genetic information so that inoculating with the nucleic acid alone cause infection and could reproduce the complete virus.
15. Kassais (1962) was the first to describe the satellite virus (Sv) which has been found only in association with tobacco necrosis virus.

16. Doi et.al. (1967) recognised MLO disease (Yellow witches broom).
17. Ishiie et.al. (1967) showed that the MLO bodies and the symptoms disappeared temporarily when the plants were treated with tetracycline antibodies.
18. Black and Markhan (1963); Miura et.al. (1966) showed that wound tamout and rice dwarf viruses contain double started RNA.
19. Shephard et.al. (1968) showed that cauliflower mosaic virus contains double stranded DNA.
20. In 1971, Diener determined that the potato spindle tuber disease was caused by a small (250- 400 bases long), single stranded circular molecule of infectious RNA which he called a viriod.
21. Windsor and Black (1972) observed RLO in the phloems of clover plants infected with clu leaf disease.
22. Davis and Worley (1973) observed motile; helicle microorganism associated with corn stunt disease and named it spiroplasma.

Economic Importance of Plant Viruses

Viruses cause serious disease of crop plants reducing both quality and quantity of final produce. Also keeping quality of plant produce is also affected. Some of the important viral diseases are:

Tobacco mosaic virus, Tomato mosaic, Tomato spotted wilt, Potato leaf roll, Potato virus X and Y in Potato, Papaya mosaic, Citrus tristiza, chilly leaf curl, Banana bunchy top etc.

Nature of Virus: Each plant virus consists of two components the nucleic acid and the protein coat or capsid. The mature particle of a plant virus is generally called virion and the whole infective particle is called as Nuclcocapsid.

Nucleic Acid of Viruses

1. Nucleic acid portion of the virus particle is called as Genome.
2. It is actual infective component.

3. Majority of plant viruses contain RNA (Ribonucleic acid).
4. Some viruses (Cauliflower mosaic virus, mung bean yellow mosaic virus, maize streak virus) contain DNA (Deoxyribo nucleic acid).
5. The nucleic acid may be in a single strand or in a double strand.
6. Most plant viruses contain single strand of RNA.

These strands may be free at both ends (liner RNA strand) or the end may be jointed together to form a circular nucleic acid molecule (Circular RNA).

7. In double stranded viruses (as RNA or ds DNA) the two strands are called around each other helically.
8. The nucleic acid may be present as a single continuous strand or it may be present as two or more piece in the same or different particles.

The Virus Protein

1. The virus protein is termed as capsid.
2. It protects nucleic acid (RNA or DNA) of virus.
3. It is made-up of different amino acid sequences in different viruses. This amino acid are linked by Peptide bonds.
4. The protein sub units of the capsid are known as structural units when these units form groups of 2,4,6 are called Morphological units or capsomers.
5. The sequence of amino acids constitutes the primary structure of the protein. Helical structure produced by amino acids is known as secondary structure of the protein and folding of protein gives tirtiory structure.
6. Protein shell or coat protects viral nucleic acid from environment.

Membrane Bound Viruses: The rhabdo such as potato yellow dwarf virus, lettuce necrotic yellows are provided with a outer envelops or membrane bearing surface projections. Inside the membrane is the nucleocapsid.

Satellite Viruses: These are viruses associated with certain typical viruses but depend on the latter for multiplication and plant infedtion and reduce the ability of the typical viruses act like parasite of the associated typical virus.

Satellite RNAs: These are small, liner RNAs found in virions of certain multicomponent viruses.

Viroids: These are small (250-400 nucleotide), naked, single stranded, circular RNAs capable of causing disease in plants by themselves.

Virusoids: These are viriod like, small, single stranded , circular RNAs that are present inside some RNA viruses, virusides are the part of genetic material of these viruses and therefore , form an obligatory association with these viruses so that neither the virus not the virusolcan multiply and infect a plant in the absence of its partner.

Morphology of Viruses

Plant viruses are usually described as:

Elongated (Rigid Rod or Flexious thread): a. Rigid rod – (Ex) Tobacco mosaic virus (15 X300 nm) Barley stripe mosaic virus (20x 10 nm). b. Flexuous thread (Ex) Potato virus X 10-13 wide upto 480 nm length , citrus tristeza virus (10-14 X 2000nm).

Rhabdo Viruses: Bacillus like- These are short bacillus like rods approximately 3 to 5 times as long as they are wide.

(Ex): Potato yellow dwarf virus 75X 380 nm.

Wheat striate mosaic virus – 65X 270 nm.

Lettuce necrotic yellow virus – 52 X 300 nm.

Spherical (Isometric or Polyhedral) : All spherical viruses are actually polyhedral ranging in diametre about 17 nm to 60 nm.

Ex: Tobacco Necrosis satellite virus 17 nm in diametre.

Wound tumour virus 60nm in diametre.

Tomato spotted wilt virus 70-80 nm in diametre.

Properties of Plant Virus

Biological Properties: Each virus produces its own protein, its function is to protect RNA from host enzymes (Ribonuclease), heat, ultra violet light and chemical protein has no infectivity.

Infectiousness of Virus

1. Viruses are infectious and highly contagious, infectivity depends on virus synthesis.
2. After the entry of pathogen in host through natural opening or wounds or pollen grains, virus comes in contact with with inoculation.
3. This is done by Host, since viruses do not produce enzymes. They lack the Lipman Enzymatic system for the conversion of high energy into potential energy required for biological activity.
4. So they have to depend on hosts. This is a major difference between the host parasite relationship in viral diseases and those of other pathogen.
5. The naked RNA induces host cell to form enzyme RNA polymerase. These enzyme in presence of viral RNA and nucleotides produce additional RNA. The new viral RNA induces host cell to produce specific protein molecule required for its coat.

Physiological Properties of Virus

(i) *Dilution End Point (DEP):* Dilution of Juices containing the virus may or may not be affected the infectivity. Tobacco mosaic virus is reported to remain infectious even in a dilution of 1:1000000 (6) Cucumber mosaic virus retains, Virulence at 1: 1000 Potatorugose mosaic virus causes poor infection when diluted to 1:10 or 1: 100.

(ii) *Thermal End Point (TEP):* Effect temperature, no of viruses can be inactivated by Thermo therpy. Tomato mosaic virus is destroyed by treating virus containing Juice at 85 to 90 degree centigrade for 10 minutes, lower temp. Seems to have no effect on the viability of plant viruses.

(iii) *Longevity in Vitro (LIV):* Retention of infectivity in storage Storability of Juice), if the Juice from mosaic infected tobacco plant is kept; the virus can remain viable for a few hours to several months. TMV in dried leaves or in Juice dried on filter paper remains infective for many years.

Chemical Properties

(i) *Host Range:* Host range is useful in distinguishing the viruses from one another. The viruses are inoculated into indicator an plant which develops typical symptoms, local lesions, ring spot systemic symptoms etc.

Ex: EMV and TMV incite symptoms on tobacco but EMV affects cucumber systematically where as TMV does not.

(ii) *Mutability and Strains:* The presence of genetic material in the form of RNA in plant viruses ensure that new strains of the viruses may develop probably by mutation of RNA. In tobacco mosaic virus alone, there are more than 50 strains.

(iii) *Serological Reactions:* If a virus containing juice is injected into body of a rabbit, the rabbits form antibodies, which will react with viral proteins to give precipitation. The reactions are specific. i. e the antibodies obtained by inoculation of strain A, of a virus will precipitate , the juice containing the same specific type of virus.

Distribution and Movement of Viruses in Plant

1. After introduction into cell, virus moves towards site of synthesis. This movement is passive through protoplasmic stream.
2. The movement of synthesised virus particles between cells occurs through plasmodesmata connecting adjacent cells.
3. The rate of spread is greater in young than old tissues. The movement faster at higher than lower temperature due to fast streaming of protoplasm. Some viruses transported through phloem, few through xylem vessels.

Virus and its Antibody are Brought together in Several Ways

(i) *Precipitation test:* In this antibodies and antigens are mixed together in solution.

(ii) *Ring interface test:* They meet at the interface between 2 solutions containing each separately.

(iii) *Gel diffusion test:* The diffuse towards each other through an agar gel and meet in a zone in suitable concentration.

(iv) *Agglutination reaction:* Sometimes antigen is observed on a surface of large particles, such as a cell, plastid and these are precipitated by addition of antibodies.

Symptoms of Plant Viruses

Definition

Symptoms: Symptoms are the expression of the diseased condition of the plant.

All most all virus disease seem to cause some degree of reduction in yield and the length of life of virus infected plants is usually shortened. The most obvious symptoms of virus infected plants are usually those appearing on the leaves but some viruses may cause striking symptoms on the stem, fruits and roots.

Symptoms: I) External (Macroscopic) :

Local: These are the symptoms produced at the site of artificial inoculation on leaves with virus.

(i) *Chlorotic Local Lessions:* Infect cell loose chlorophylls and other pigments.

Ex. TMV on Cowpea host.

(ii) *Necrotic Local Lessions:* Infect cell die.

Ex. TMV on Nicotina Glutinosa host.

(iii) *Ring Spot Local Lesions*: Conist of central group of died cells near inoculated area (Necrotic ring) Ex. Potato Virus: On Chenopodium Amaranticolar host.

Systemic: In almost all viruses of plant occurring in field, the virus is present throughout the plant.

Variegation (Colour Breaking):

1. Decolouration in patches on leaves.
2. Sharpness of boundaries.
3. Distribution and discolouration of lamina.

Mosaic: Mosaic are characterised by light green yellow or white areas intermingled with normal green of leaves of fruit or of whitish areas are intermingled with the areas of the normal colour of flowers or fruits depending upon intensity or patterns of discolouration mosaic type symptoms may be described as follows.

(a) Streak or stripe: Red stripe of jowar.

(b) Vein clearing: BCMV

(c) Vein banding: Beat curly top.

(d) Inter veinal mosaic light discolouration restricted in bet. veins.

Line Pattern

(a) Oak leaf pattern (OLP) (Apple oak leaf)

(b) Systemic Rings (AMV on tobacco)

Mottling: If the discoloured patch of a variegated leaves are rounded the variegation is usually designated as mottling.

Malformation

1. Change in leaf form, any deviation from normal.
2. This includes uneven growth of leaf lamina leaves becomes curled, britle (Crinkling) and show, prominaces and depressions (puckering) upward and downward curling, vein distortion, leaf enation, galls and tumours.

Others

(a) Reddening of leaves cotton red leaf due to abnormal accumulation of anthoecynamin.

(b) Blackening of veins Potato virus 'Y' in potato (Due to more synthesis of melanis).

(c) *Bronzening:* Ex. TSWV, in Tomato

(d) *Etching:* Tobacco etch virus.

Internal (Microscopic)

These include inclusion bodies. Inclusions are microscopic bodies produced by some of the viruses. These are produced in cytoplasm or nucleic acid.

Cytoplasm

1. Amorphous or amoeboid also called as X-bodies produced by spherical or oval viruses Ex. CMV.

2. ***Crystalline:*** These type of inclusions are produced by rod shaped viruses. Ex. Red clover vein mosaic, cactus virus, and petunia ring spot.

Nucleus

Ex. Tobacco etch virus, rectangular plates. BCMV- Strain pisum Virus -2. Isometric crystals.

Proliferation (Toratonic Symptoms)

1. Asymmetry of leaf lamina Ex. Grape fan leaf.
2. *Blistering:* Dark green area may be raised to give blistering effect. Margin of leaves twisted ex. BCMV.
3. *Swelling of stem:* Cocoa swollen shoot virus.
4. *Flattening of branches and distribution of stem:* Apple flat limb.
5. *Enations:* These are the out growth in different shape and form either on veins or leaf lamina.
6. *Vein enation:* Citrus onation virus, pea enation mosaic.
7. *Leaf enation:* Cotton leaf curl: Tobacco leaf curl;
8. *Tumours:* Clover wound tumour virus.
9. *Galls:* Woody galls of citrus.
10. *Pitting:* A) Stem pitting – Apple stem pitting, citrus twisteza virus on kagzi lime.

    B) *Fruit pitting:* Pear stony pit- Apple groove virus on apple.
11. *Stunt:* Tomato bushy stunt.
12. *Dwarfing:* Barley yellow dwarf.
13. *Leaf Roll:* Potato leaf roll.
14. *Yellows:* Beet Yellows.
15. *Pox:* Plum pox virus.

Definition of Latent Viruses and Masked Symptoms

Latent Viruses: Many viruses may infect certain hosts without ever causing development of visible symptoms on them. Such viruses are usually called as “Latent viruses” and hosts are called as “Symptomiess carriers”.

Masked Symptoms: Virus induced plant symptoms that are absent under certain environmental conditions, but appear when the host is exposed to dertain conditions of light and temperature.

Transmission of Plant Viruses

Plant viruses are not disseminated as such by wind or water. Viruses are transmitted from plant to plant in a number of ways such as vegetative, propagation, mechanically through sap and by seed, pollen, insect, mites, nematodes, dodder and fungi.

Transmission by Vegetative Propagation: Plants are propagated vegetatively by budding or grafting or by cutting or by the use of tubers, corms, bulbs or rhizome.

Any virus present in the mother plant from which these organs are taken will almost always be transmitted to the progeny. Transmission of viruses may also occur through natural root grafts of adjacent plant.

Mechanical Transmissions: Such transmissions may takes place between closely spaced plants after a strong wind by contact , when plants are wounded during cultural operations, virus infected sap adhering to the tools, worker hands or cloth accidently transmitted to the subsequently wounded plants.

E .g. Potato virus: (PV-X,TMV on Tobacco and Tomato)

Natural Modes of Transmission: This includes air borne transmission through insects and mites and soil borne transmission through nematodes and fungus.

Seed Transmission: About 100 viruses have been reported to be transmitted by seed only a small portion (1-30) of the seed derived from virus infected.

Plant transmit the virus. Frequencies varies with the host virus combination. Tobacco ring spot virus is soybean may be transmitted by almost all 100% of the seed of infected plant. 28-94% in squash mosaic virus in musk melon. 50-100% in barley stripe mosaic virus in barley. Mosaic of Datura stramonium to the tune of 79%.

Pollen Transmission: Virus transmitted by pollen may infect not only the seed but also seedling that will grow form it. Ex. Tomato and Tobacco black ring virus.

Dodder Transmission: Several plant virus can be transmit ted from one plant to another plant through the bridge formed between the two plants by twining the stem of parasitic plant, dodder (Cuscuta species), green strain of cucumber mosaic virus on N. glutinosa, lucern (alfa-alfa) , mosaic virus to tobacco potato stem mottle virus to tobacco. Bennett (1940) showed dodder transmission, sugar beet curly top virus (BCTV), cucumber mosaic virus.

Natural Modes of Transmission of Viruses

This includes air borne transmission through insects and mites and soil borne transmission through nematodes and fungus.

Air Borne through Insects: The most common and economically most important means of virus transmission in the field. Members of the order Homoptera- Aphids, Jassids, Leaf hopper, white flies, mealy bug , scale insects. Thysanoptera – Thrips, coleopteran- Beetles. Insects with sucking mouth parts carry plant viruses on their stylet- stylet borne or non persistances.

Circulative

Circulative Viruses (Persistent): Viruses that are acquired by their vectors through their Mouth Parts, accumulate internally, then they are passed through their tissues and introduced into the plants again via mouth parts of the vectors.

Propagative Viruses: Some Circulative viruses may multiply in their respective vector are then called Propagative viruses.

Insect Vectors: Virus transmitting insect is called vector Walker (1969) showed that viruses are mostly really on insect for transmission (400 species of insect vector transmitting more than 200 viruses). The viruses are transmitted by different types of insects and most efficient vectorsare of sucking and bitting type of insects.

Group of Viruses based upon the Action of Vector

1. Non persistent viruses
2. Persistant Viruses
3. Latent viruses
4. Bacteriophage
5. Circulative viruses

Aphids: These are the most important insect vector of plant viruses and transmit great majority (about 170) of the all stylet borne viruses Ex.

Soybean mosaic, cowpea aphids, borne mosaic, potato leaf roll, pea enation mosaic, lettuce necrotic yellows, red clover mosaic BYMV, MCMV, AMV.

White Flies: Tobacco leaf curl, yellow vein of okra, pumpin yellow mosaic, mung yellow mosaic, cotton leaf crumple, cassava mosaic, sweet potato mosaic.

Leaf Hoppers: At least 40 plant viruses are transmitted by leaf hoppers, all leaf hoppers transmitted viruses are circulatory, several are known to multiply in the Vector (Propagative), cause disturbance an phloem region ex. Rice, dwarf viruses, Maize mosaic, Rice tungro viruses, Beet curly top, Potato yellow dwarf, Maize Rough dwarf.

Thrips: Tomato spotted wilt virus.

Beetles: Cowpea mosaic virus, cowpea chlorotic motile virus, bean pod mottle, squash mosaic, raddish mosaic.

Mites: Transmit viruses like fig mosaic, tur sterility mosaic, peach mosaic, wheat streak mosaic.

Soil Borne (Nepo Viruses) Nematodes

NEPO: Nematode transmitted polyhedral soil viruses.

Ex. Tobacco ring spot viruses.

NETU: Nematode transmitted tubular soil viruses- tobacco.

Soil Borne Nematodes: Approximately 20 plant viruses have been shown to be transmitted by one or more species of 4 genera of soil inhabiting Ectoparasitic Nematodes.

Nematode Virus Vectors	***Example***
Xiphinema index	Grape vine fan leaf
Xiphinema coxy	Cherry leaf roll
Xiphinema americaum	Ring spot of tobacco, tomatomsoybean bud blight, bringal mosaic.
Longidours	Rasberry ring spot, tomato black ring.
Trichodours and pare trichodorous	Rod shaped viruses, tobacco rattle pea early browing.

Fungus: Root infecting fungus-Olpidium transmits 4 plants viruses.

V Fungus Virus Vectors	*Example*
Olpidium brassicae	Tobacco necrosis , cucumeber, necrosis (O.Cucubitacearum) lettuce big vein tobacco stunt virus
Polymyxe graminis	Wheat mosaic virus, beat necrotic yellow vein, virus.
Spon Gospora subteranea	Potato mop top virus
Synchytrium endobioticum	PU-X virus

Control of Plant Viruses

Quarantine Laws: The best way to control virus disease is to keep it out of an area through a system of quarantine, inspections and certifications.

Selection of Seed: Select seed from disease free localities.

Selection of Planting Materials: i. e Cutting, bull, rhizomes, tubers etc should be free from disease.

Eradication: Eradication of diseased plant to eliminate the inoculum from the field.

Indexing: Periodical indexing of the mother plants, producing Propagative organs is necessary to ascertain these continuous freedom from virus.

Protection against Insect Vectors: This can be done by growing trap crops to check the insect vectors. Ex. Cotton reddening, white flies in bhendi.

Soil Fumigation: Nematodes transmitted viruses can be reduced by the soil fumigation to control nematodes.

Weeds: Removal and destruction of weeds that serve as host, broad leaf weeds in banana, orchard reduce bunchy tops.

Use of Resistant Varieties: Ex. Parbhani Kranti, yellow vein mosaic of the bhendi, sugarcane mosaic.

Immunization: The disease caused by severe strains of virus can be avoided if the plants are inoculated first with a mild strain of some viruses. E.g. Citrus greening.

Temperature Treatment: Sugarcane mosaic can be destroyed or reduce by hot water treatment 520 C for 30 minutes.

Use of Insecticides: Control of insect vector with the use of insecticides.

Non Availability of Chemical Substances: (Demerits): No chemical substance (Viroides) are yet available for controlling virus disease of plants in the field.

Mycoplasma like Organisms or Pleuro Pneumonia like Organisms (PPLO)

Mycoplasmas are the smallest, walless, free living microorganisms, cause disease similar to virus in plants. Doi ct.al. (1967) described the structure of this pathogen using election microscope. Mycoplasmas like bodies have also been detected in electron micro graphs of plants suffering from corn. Stunt and aster yellow.

Definition: "Mycoplasma are smallest, free living, walless, prokaryotic organism without an organised and bounded nucleus." Mycoplasma lack rigid cell wall, being surrounded only by single tripla layered unit membrane which allows them to be highly plemorphic.

They assume vast array of shapes and size. They require sterols (Lipoproteins) for growth Mycoplasma cannot be grown on artificial media and they reproduce by budding and binary fissions.

Taxonomy: Mycoplasma like organisms are included in class: Mollicutes which has one order: Mycoplasmatales. The order has three families, each family containing one genus.

1. *Mycoplasmataceae:* One genus Mycoplasma.
2. *Acholeplamataceae:* One genus Acholeoplasma.
3. *Spiroplasmataceae:* One genus Spiroplasma.

These Genera are Separated on the Basis of their Sterol Requirement for Growth

Characteristics of Mycoplasma like Organisms (MLOs)

1. MLOc lack true cell wall and ability to synthesize the substances required to form a cell wall.
2. MLos are bounded only by a single triple layered unit membrane.

3. They are small, sometime ultra-microscopic cell containing micro cytoplasm, randomly distributed ribosome and strands of nuclear material.
4. They measure from 175 to 250 nm in diametre during reproduction but grow into various shapes and size lateron.
5. Shapes range from coccid or slightly ovoid to filamentous, sometimes, they produce branched myceloid structure.
6. They are capable of reproducing by budding and binary transverse fission on coccid and filamentous cell.
7. MLOs have no flagella, produce no spore and are gramve.
8. Nearly all MLOs are parasitic to humans and animals and all saprophytic.

   Once can be grown on more or less complex, Artificial nutrient media in which they produce minute, colonies, that usually have a characteristic "fried egg" appearance.
9. MLOs have been isolated mostly from healthy and diseased animals and human suffering from the disease of respiratory and urogential tracts (associated with arthritic and nervous disorder of animals).
10. Most MLOs are completry resistance to penicillin, but they are sensitive to tetracyclic, chloramphenicol, some to erythromycin and to certain other anti biotics.

Mycoplasma like Organism of Plant

1. The organism observed in plants and insect vector with the exception of Spiro plasma, resemble the mycoplasmas of the genera, Mycoplasma in all morphological aspects (lack cell wall bounded by unit membrane have cytoplasm ribosomes and strands of nuclear material. Their shape is usually spherical to avoid or irregularly tubular to filamentous and their sizes comparable to those of the typical Mycoplasma.
   (a) Plant Mycoplasma like organism are generally present in the sap of a small number of phloem sieve tubes.

(b) Most plant Mycoplasma like organisms are transmitted from plant to plant by leaf hopper but some are transmitted by psyllids and plant hoppers.

(c) Plant Mycoplasma like bodies also grow in the alimentary panals, hemolymph salivary glands and intracellularly in the various body organs of their insect vectors.

(d) The vector cannot transmit the Mycoplasma, immediately after feeding on infected plants, but it begins to transmit the Mycoplasma, immediately after feeding on depending upon temperature.

Shorter incubation period occurs at 30 0 C where as longest at about 10 0C.

The incubation period is required for multiplication and distribution of the Mycoplasms within the insect. If Mycoplasma is acquired from the plant, it multiplies first in the intestinal cells of vector, it then passes into the hemolymph and internal organs are infected.

When the concentration of mycoplasms in salivary glands reachs to a certain level, the insect begins to transmit the pathogen to new plant and continuous to do so for the rest of its life.

MLOs are usually be acquired readily or better by nymph than adult hopper and survive through subsequent molts, but are not passed on the adult to the egg and to next generation.

Symptoms Produced by Mycoplasma like Organisms (MLOs) in Plants

1. Mycoplasma like bodies are now stated to be occur in more than 60 to 70 plant diseases, which are characterised by the growth abnormalities and yellowing of leaves.
2. Characteristic symptom of yellow type disease includes uniform yellowing or reddening of leaves, smaller leaves, shortening of internodes, stunting of plants and proliferation of auxiliary bud.
3. "Witch brooms" includes reduction of leaf size with leaves becoming brittle, excessive proliferation of shoots.

4. "Phyllody" replacement of floral parts of leaves greening or sterility of flowers and reduced yield. Finally more or less rapid dieback, decline and disorder after several years.
5. MLOs are mostly restricted to phloem tissue becomes; a phloem may provide favourable conditions for growth as phloem elements have a high osmotic pressure and slightly alkaline PH.

Transmission of Mycoplasms like Organisms (MLOs)

1. MLOs are found exclusively in phloem tissue of infected plants and require a vector for pheir dispersal.
2. A majority of MLOs are transmitted by leaf hopper, tree hoppers, plant hopper, aphids, and mites also play a role in transmission of MLOs.
3. The vector acquires the MLOs while feeding on contaminated phloem. Incubation period ranges between 10 to 45 days depending upon the temperature. During this period MLOs multiply in the body organs of vectors particularly in salivary glands and hemolymph.
4. MLOs are not passed on through the eggs of contaminated insects.
5. In addition to vectors, MLOs are also transmitted by grafting and by the plant parasite like dodder.

Disease Caused by Mycoplasma like Organisms (MLOs)

1. Aster yellow
2. Lethal yellowing of coconut
3. Elm phloem necrosis (Elm yellow)
4. Peach (X-Disease)
5. Pear decline
6. Citrus stubborur
7. Corn stunt
8. Sandle spike
9. Citrus greening
10. Seasamum phyllody

11. Grassy shoot of sugarcane
12. Little leaf of Brinjal
13. Rice yellow dwarf.

General Control Measures of Mycoplasma like Organisms

1. Most of the control measures against yellow type of disease are avoided at preventing infections.
2. *Use of antibiotics:* Antibiotics are known to suppress the yellow type disease. The most effective antibiotics appear to be chloro tetracycling, oxy tetra cycline and tetra cycline with choramphenicol, ledermycin and metha cycline. The application of antibiotics by root dip or paste under the bark, standing cutting in solution appears to be more effective than foliar spray or soil drenches.
3. *Heat Treatment:* It is possible to cure yellow infected plants grassy, shoot of sugarcane by heat treatment.
4. *Vector Control:* Since most of disease are spread by the vector, control of vector through effective insecticides like rogar, Malathion, Endosulphan, demecron etc.
5. *Kiliing of alternate hosts:* Weeds may serve as alternate host, hence killing of alternate host.

Chapter 6

Factors Responsible for Non Parasitic Disease

Unfavourable Tcmperature

High Temperature Effects

1. Plants are generally injured faster and to a greater extent when temperature becomes higher than maximum for growth.
2. High temperature are usually responsible for sun solid injuries appearing the sun exposed sides of fleshy fruits and vegetables.

Ex. Sun scald of apple, peppers, tomato, onion-bulbs, potato tubers, canker of linseeds.

Low Temperature Effects

1. Low temperature causes greater damage to crops than high temperature.
2. Blotch type necrosis in Potato is due to freezing injury.
3. The frost injuries involved killing of buds of peach, cherry, killing of flowers, young fruits and sometimes succulent twings of most trees.
4. Low winter temperature may kill young roots of trees such as apple and may also cause bark splitting and canker development.

Unfavourable Light: Lack of sufficient light or absence of light retards chlorophyll formation and promotes slender growths with long internodes, thus leading to pale green leaves, spridly growth and premature drop of leaves and flower. This condition is known as "Etiloation".

Moisture Effects

Low Moisture Effects: Plant suffering from lack of sufficient soil moisture usually remain stunted are pale green to light yellow, have few small dropping leaves and finally in absence of moisture plant dry and wilt.

High Moisture Effects: Stagnation of water or excessive moisture by flooding in the field may cause decay or rotting of fibrous root of plants resulting into wilting. Primarily due to accumulation of toxic materials around the root and base of stem and also due to non availability of nutrients.

Ex. Tip burn of Paddy

Another common symptom of house plants caused by excessive moisture is swelling. Edema appears as small bumps on the lower of leaves or on stem.

Inadequate Oxygen: When there is excessive respiration in closed atmosphere the entire oxygen supply may be exhausted resulting in disintegration of cells due to enzymic action.

Ex. Black Heart of Potato

Atmospheric Impurities or Air Pollution: Presence of injuries gases in the atmosphere may cause definite injury to plant and plant parts.

More serious and wide spread damage is caused to plant in the fields by chemicals such as Ozone, sulphur dioxide hydrogen fluoride, nitrogen dioxide, peroxyaacotyl nitrates.

Ex. Black tip of mango necrosis. Fruit on trees is close proximity to brick kilns may bear necrotic lesions and become useless for sale and consumption. The smoke of kills polluted the air with toxic gases like gases like sulphur dioxide which caused necrosis of tissues.

Toxic Effects of Decomposition Organic Matter in Soil and Soil Mineral Toxicity: Crop residue decomposing in soil produce toxic substances such as fatty acids which produce symptoms of damping off, root rot, wilt and nutritional deficiency.

Excessive amount of sodium salt especially sodium chloride, sodium sulphate and sodium carbonate raise the PH of the soil and cause alkali injury i.e Chlorosis, stunting, boron, manganese and copper have been most frequently implicated in mineral toxicity disease. Excess boron is toxic to many vegetable and trees, Excess manganese is known to cause crinkle leaf disease in cotton.

Herbicidal Injury: Some of the most frequent plant disorder seems to be the result of extensive use of herbicides. Increasing number of herbicides in use for general or specific weed control has created problems.

Nutritional Deficiencies or Disorder in Plants

(i) Deficiency of minerals viz Nitrogen, Phosphorus, potash, manganese , magnesium, boron, zinc, copper , iron, etc. results in disorders in plant metabolism and cause hunger signs in the crops.

(ii) Excess of minerals disturbs nutritional balance needed for good metabolism in the palnt, thus hindering the effect of essential element.

(iii) The deficiencies and excess of minerals also reduce the resistance of plant to fungal, bacterial and other diseases.

(iv) The kind of symptoms produced by deficiency of a certain nutrient depend primarily on the functions of that particular element in the plant.

Plant Disease due to Lack of Minerals

Other Improper Agricultural Practises: A variety of other agricultural practices improperly carried out, many cause consideration damage to plants. Improper application of chemical viz. fungicides insecticides, nematicides and fertilizers at too high concentration results, into losses. Spray injury results into leaf burn or spolting or rosseting of fruits. Excessive or deep cultivation between rows of growing plants may pull roots of many plants.

Flowering Plant of Parasites and Its Classification

1. Although most of the plant disease are caused by fungi, bacteria, viruses and nematodes. There are few seed plants, which are parasitic on living plants.
2. In many cases the damage caused by these parasites is slight on attacked host plants are little importance,

but there are some examples where those flowering plants attack valuable crops and trees causing considerable losses.

3. Some of these attack on root while others parasitize stem, some are devoid of chlorophyll and entirely depended on their hosts for food supply while other have chlorophyll and obtain only mineral constituents of food from the host.
4. More than 2500 species of higher plants are known to live parasitically on other plants.

Classification of Flowering Plant Parasites

(i) Complete Parasite (Holoparasites):

(a) Root Orobanchea (Broom rape)

(b) Stem Cuscuta (Dodder Amarvel)

(ii) Partial Parasites (Semi Parasites)

(a) Root Striga (Sandle wood witch weed)

(b) Stem Loranthus (Banda or Deudrophae)

Complete Stem Parasite - Dodder

Family: Cuscutaceae

Genus: Cuscuta

1. These are non-chlorophyll bearing leafless, twining parasitic seed plants. They are yellow, pink or orange in colour, they attach to the host.
2. The first appearance of parasite in field is noticed as small masses of branched, thread like, leafless stem, which are devoid of green pigment and twice around the stem or leaves of the hosts.
3. The leaves are represented by minutes functionless scales, which are evident on close examination, when the stem of parasites comes in contact with the hosts, the minute root like organs, haustoria penetrates into the host cortex and serves as an organ of food absorption.
4. When the relationship with the host is firmly established, the dodder plant losses the contact from soil.

5. The flowers are found in clusters, they are tiny, white, pink or yellowing in colour. The seeds are formed in capsule. A single plant may produce as many as 3000 seeds.
6. Dodder may also serves as a bridge for transmission of viruses from virus infected to virus free plant. There are 3 species of dodder:

   i) *Larger seed:* - Cuscuta indicora

   ii) *Small seed:* - Cuscuta planiflora

   iii) *Field dodder:*- Cuscuta campestris
7. The common dodder Cuscuta granovil attacks clovers, berseem flag and many other oilseed crops. It also attacks ornamental and hedge plants. It is fast developing parasites and within 2-3 seasons may destroy a complete plant.
8. The dodder perepetuates through seeds which remain dormant in the soil unitl favourable seasons returns, stem portion of parasite is also meant of perpetuation.

Control Measures of Dodder

1. Selection of dodder free seeds for sowing.
2. Preventing the movement of grazing animals from infected field to clean field.
3. Restriction of flow of irrigation water through infected field.
4. Destruction of flow of irrigation water through infected field.
5. Crop rotation with non host plants.
6. Dodder can be controlled by the use of soil herbicides, such as chloroproopham, DCPA, Dichobhil, Donoseb, glyphosate, these chemicals kills the dodder plant upon its germiantion from seed.

Complete Root Parasite - Broom Rape

Family: Orobanchaceae

Genus: Orobanche.

1. In some areas of the world, broom rape cause losses varying from 15 to 70 % of the crop, Orobanche spp. are total root parasites affecting tobacco, Brinjal,

Tomato, Cauliflower, Turnip and many other solanaceous and cruciferous plants.

2. It is especially destructive to tobacco and Brinjal. The parasite consists of a stount, fleshy stem 15 to 50 cm long. This stem is yellow or brownish red in colour and is covered by small thin and brown scaly leaves.
3. The flower appearing in the axil of the leaves are white and tubular seeds are very small and black in colour and many remain viable in soil for several years.
4. The haustoria of the parasite penetrate into the root of the hosts and draw it nourishment. When the host is carefully uprooted, the parasitic roots are seen intertwined with the host root systems; the growth of the host is retarded and remains stunted.

Control Measures of Broom Rape

1. Destroy the parasite before flowering.
2. Long crop rotation.
3. Spraying the soil with 0.25% copper sulphate (drenching) solution has been reported to be successful in destroying the parasites.
4. Fumigating the soil with methyl bromide.
5. It was recently reported that broom tapes were effectively control after treatment with the herbicides "Glyphosate".

Partial Stem Parasite - Loranthus or Bandgul

1. It is common parasites of mango trees. In Northern India 60-90% of mango trees and large no of other trees are heavily or moderately infected by these parasites.
2. Loranthus Dendro-phthae falcate, the most common species in India is the semi parasitic of the tree trunk and branches. Their leaves posses chlorophyll and synthesize carbohydrate constituent of their food requirement.
3. Since the parasite attacks the aerial part of host tree, situated far above the soil level and as such devoid of root system of its own.

4. The parasite attacks, the aerial parts of the host trees, by developing haustoria and obtain its nourishment directly from the vascular system of the host plant.
5. The continuous sucking of the food material by parasite resulting the host to die.
6. The place at which host is attached and where the haustoria penetrates often swell to form tumours which vary in size according to age of the parasites.
7. The flowers of parasites are borne in clusters. They are long tabular in shape and usually greenish, red or white in colour according to species.
8. The fruit is fleshy and contains a solitary seed. It is sweet eaten by birds and animals.
9. The parasite is spread by dispersal of its seed, mostly through birds and to some extent by other animals.
10. The damage done by the parasite is most marked in production of new growth of the host. The quantity and yield of fruit is considerably lowered. Leaves are reduced in size and show unhealthy green colour.

Control Measures of Loranthus

1. Scrapping of parasite before seedling from the infected bunches.
2. Sowing of branches sufficient low to the tumours.
3. Injection of CuSo4 or 2-4-D into the infected branches has been found effectively in eradicating the parasite from mango.

Partial Root Parasite - Striga

Family: Scrophulariaceae

Genus: Striga witch weed (Striga species) is well known partial root parasite of sugarcane Jawar, maize, cereals and millets in India. There are four species of Striga reports in the country on sugarcane, rice , sorghum and other millets.

1. S.densiflora
2. S. quphrasioldes
3. S.lutee

4. S.asiatica.
2. These plants although obligate parasites do not obtain all of their nutrient material from their host root. They posses' chlorophyll bearing leaves. Plants are 20-60 CM in roots of cereals, sugarcane, etc.
3. Striga can be found on light as well as heavy soil in rabi and kharif season. The seeds of Striga are very minute and produced in great abundance 50,000 to 1, 00,000 seeds /pl/yr.
4. One flower / capsule contains 1200-1500 seeds. Viability of those seeds has been reported to be from 12-40 years. Short distance dissemination of these seeds.
5. For germination of seeds of Striga species, stimulant provided by the root executes of specific host is essential.
6. Seed starts germination after 7-10 days. In most cases germination occurs 40 cm depth. After germination, the parasites grow below the soil surface for about 4-8 weeks and produces underground stem and root.
7. The underground portion of the stem contains bud in the axil of leaf. Stem of parasite forms haustoria which penetrates the root of hosts plants and also water and nutrient eventually wasting and destroying the host.

Striga Difficult to Control Because

1. Heavy seed production 50,000 to1, 00,000 per plant.
2. Seed viability is longer.
3. Most hyyries are stimulating positive.
4. Alternative host like grasses are present on bunds.

Control Measures of Striga

1. Complete eradication of parasite before flowering.
2. Regular Interculture should be followed.
3. Crop rotation with cotton-Jowar- Groundnut.
4. Weedicides are used to control Striga before flowering.
   (a) 2-4-D @ 2.5 lit/500 lit of H2O per ha.
   (b) Attrazine @ 2 kg / 500 lit of water per ha.

(c) 1% TCPA @ 45 kg/ha.

(d) Deep ploughing after harvest reduces the vialibility of seed.

(e) Fallowing.

Post Harvest Management of Fruits and Vegetables

Introduction to Post Harvest Management of Fruits and Vegetables: Improper harvesting, handling transportation and distribution of fruits and vegetables result in the significant losses which cause ultimately economic loss. The reduction of post-harvest losses reduces the dependence on imports of commodity fertilizers pesticides and other chemicals save a substantial amount of foreign exchange. It is estimated that total loss of vegetable and fruits in India due to inadequate post-harvest handling transportation of storage at less 20-25%.

The importance of reduction in post-harvest loss of fruit and vegetable is of vital importance countries like India, Malaysia and knowledge of post-harvest management and loss of fruits and vegetables. Several factors influence the post-harvest losses due to physical, physiological, mechanical and hygienic conditions. Fruits and vegetables are characterized by high metabolic activities and known to possess short shelf life. Recent development has improved shelf life of fruit and vegetables.

Maturity of Fruits and Vegetables

It refers to the attachment of final stage of biological function by a plant part or plant as a whole.

Or

It is the particular stage in life of plant of fruit at which they attain maximum growth and size.

Good quality of fruits and vegetables are obtained when harvesting is done at the proper stage of maturity. It is the stage where any organ of the plant attains full growth and development. So it is the stage of fruit development beyond which no further growth take place. After maturity of any organ it starts its decline stage i.e. called as "Ripening". Earlier the harvest, longer is the time of ripen. Greater the maturity, lesser are the number of days required for the fruit ripen. But the ripe fruits from early harvests and poor quality indicated by lower organoleptic ratings and with increasing maturity, quality improved. The maturity indices are also called as “Maturity Standards” or “Signs of Maturity”. Maturity signs help in judging maturity of fruits and vegetables. The signs are based on experience and skill and judgement. As the market value depends upon quality of the produce, the knowledge regarding maturity indices of right stage of harvest caries vital importance. Secondly shelf life of the produce in some fruits depends upon maturity stage of harvested produce.

There are Five Types of Indices to Judge the Maturity of the Fruit

1. Visual means
2. Physical means
3. Chemical analysis
4. Computation
5. Physiological method.

Visual Means: Skin colour, size, persistence of style portion, drying of outer leaf, drying whole plant body, change in smell or flavour, dropping down of ripe fruits.

Physical Means: Fairness easy separation or abscission, specific gravity, weight of the fruit.

Chemical Analysis: T.S.S , acids, Starch, sugar, etc.

Computation: Days for harvesting fruits from fruit set till maturity.

Physiological Method: Respiration rate, internal ethylene evolution.

Harvesting of Fruits and Vegetables

Harvesting is one of the important operations, that decide the quality as well as storage life of produce and helps in preventing huge losses of fruits. Harvesting of fruits should be done at optimum stage of maturity. During harvesting operation, a high standard of field hygiene should be maintained. It should be done carefully at proper time without damaging the fruits. The harvesting operation includes.

(i) Identification and judging the maturity of fruits.

(ii) Selection of mature fruits.

(iii) Detaching or separating of the fruits from tree, and

(iv) Collection of matured fruits.

Method of Harvesting

Different kinds of fruit and vegetables require different methods after harvesting. The methods of harvesting are:

1. Manual Harvesting
2. Mechanical Harvesting

Manual Harvesting: Harvesting by one's own hand is called manual harvesting. It is done in several ways:

(a) Ladder / bag picking method

(b) Poles/ Clippers method

(c) Harvesting by means of cutting knives

(d) Harvesting by means of digging tools.

Mechanical Harvesting: In this method numbers of mechanical devices are used for harvesting the produce on commercial scale.

Types of Maturity of Fruits and Vegetables

(a) Harvesting Maturity

(b) Physiological Maturity

(c) Commercial or Horticultural Maturity

Harvesting Maturity: The harvest maturity of vegetable depends upon the purposes for which it is harvested. For local market and for processing, fully coloured tomato fruits are harvested. However, for a distant market fruit which have started developing colour are harvested. The post-harvest quality and storage life of fruit appear to be controlled by the maturity. If the fruits are harvested at a proper stage of the maturity the quality of fruit is excellent. Poor quality and uneven ripening are caused by early harvesting and late harvesting result in extremely poor shelf life. It is imperative that the fruit should be at the right stage of the maturity with no of physical damage. Various Maturity Indices such as number of damage from fruit set, visual indicators, size, shape, colour, appearances, texture, lenticel number, specific gravity starch contain soluble solids, sugar, acid ratio and oil content are used to determine maturity of fruits.

Physiological Maturity

1. In a physiological sense, maturity refers to the attainment of final stage of biological function by a plant part of plant as a whole.

Example: A French bean pod of okra is at its physiological maturity when seeds are fully developed and the pod is which will dehisce with little pressure.

2. It is the stage of development of any plant organ at its maximum size and growth.
3. The physiological maturity is judged of measure by rate of respiration and sugar: acid ratio.
4. It always followed by senescence.

Commercial or Horticultural Maturity: It is the stage of development, when plant or plant part possesses the prerequisites the utilization by consumer for particular purpose.

1. It is a stage of fruit and vegetable at which consumer wants the fruit and vegetable or fruit and vegetable require by market. The horticultural maturity of fruits and vegetables depends upon the purpose for which it is harvested.

Example: The French bean pod of okra pod is matured when it is tender with maximum size, as per horticultural maturity.

2. It is a stage appears at any time between development and growth of any plant upon fruit.
3. It can be predicted by using different terminology like premature, mature and over mature.
4. There is no necessity of senescence.

Matured Fruits are having Completed Natural Growth and Development

Climateric Fruits	***Non- climateric Fruits***
Apricot, Almond, Banana, Custard apple, fig, guava, Mango, Sapota, Papaya, Muskmelon, Phalsa, Tomato, Watermelon.	Cashew, Cherry, Cucumber, Brinjal, Grapes, Ber, Lemon, Oranges, Pomegranate, Pepper , Cocoa.

Maturity Indices and Harvest Indices

Maturity can be described as the attainment of the particular size or stage after which ripening takes place.

On the other hand, ripening means the qualitative changes in fruits after maturity of which it become edible.

Various Maturity Indices are

(i) Number of days from fruit set,

(ii) Visual indicators,

(iii) Size of fruits,

(iv) Shape of fruit,

(v) Colour of fruit,

(vi) Appearance (External)

(vii) Texture

(viii) Lenticel number

(ix) Specific gravity

(x) Starch Content

(xi) Soluble solids

(xii) Sugar acid ratio

(xiii) Oil content

These indices are used to determination of fruits, Bhatnagar and Suramanyam (1971) reported that in Mango varieties – Alphanso and Pairi it took 110 to 125 days after fruit set for surface colour to change from dark green to olive green and flesh colour from white to pale yellow.

Haden mangos are ready for picking at 1.02 specific gravity. The general practice mango observed in this process is known as TAPKA.

The stage of maturity of Banana is determined by experience and judged largely by the visual appearances of the hanging branch and particularly by angularity of the individual banana fingers.

Harvest Indices: The quality of fruits and vegetable cannot be improved but it can be presented when harvesting is done at proper stage of maturity. Immature fruits when harvested will give poor quality and erratic ripening.

In some cases, if the produce is to shipped to distant markets, or stored, to wait for a better price, it should be picked in the mature but unripe stage. Here lies the difficulty, because unlike the ripening stage, the boundary between pre maturation and maturation stage is hard to detect. No prominent changes in firmness or colour are evident often harvest indices becomes allitrary and subjective.

Maturity Signs and Harvesting of Banana

Maturity Signs of Banana

1. The dwarf bananas are ready for harvest within 11 to 14 months after planting while tall varieties take about 14 to 16 months to harvest.
2. Fruits usually mature in 120 to 140 days after flowering.
3. The fruit bunch is harvested when the ridges on their surface changes from angular to round.
4. The dried parts of flowers at the top of fruit drop off easily.
5. The top most leaf starts drying as the bunch matures.
6. Colour of fruits or fingers changes from dark green to pale green.

Harvesting of Banana

1. The trunk is lopped with a sickle or hatchet over half-way through the stem.
2. The bunch will not fall to the ground but hang on, and injury is avoided. The bunch is held and its penduncle is served.
3. About 30 cm of the stalk must be left to make handling easy.

Maturity Signs and Harvesting of Mango

Maturity Signs of Mango

1. Fruits generally require 95 to 115 days to mature after flowering.
2. Building of shoulders and a slight depression near the stalk end indicates the maturity of fruit.

3. Colour of fruit changes from dark green to pale green.
4. Red blush develops on the fruit shoulders.
5. One or two or three ripe fruits fall from the plant naturally, indicating maturity of the other fruit on the tree.
6. The fruits with specific gravity between 1.00 to 1.02 are said to be mature.

Harvesting of Mango

1. Harvesting should be done by hand as much as possible. The fruit is twisted sharply sidewise or upward.
2. The pedicel is trimmed short enough to keep it from puncturing other fruits. Mango harvester harvests fruit borne on high branches.

Maturity Signs and Harvesting of Coconut

Maturity Signs of Coconut

1. Coconut usually matures in about 350 to 375 days after appearance of the inflorescence.
2. Colour of fruit changes from green to yellowish or brownish green.
3. The fruit produces peculiar metabolic sound on thumping.
4. All nuts in a bunch mature at the same time and uniformity.

Harvesting of Coconut

1. In India climbing the trees does harvesting. The climber often uses a small ladder. After climbing top of the ladder he uses a rope ring round the feet for climbing the rest of the tree. On reaching the crown he harvest matures nuts.
2. In Srilanka, harvesting is done from the ground with the help of a knife attached to a long bamboo pole.
3. In Malaysia and Thailand trained monkeys are also used for harvesting.
4. In Africa, nuts are allowed to fall from bunches.

Maturity Signs and Harvesting of Arecanut

Maturity Signs of Arecanut

1. Fruits is ready for harvesting in about 180 to 200 days from flowering.
2. First colour changes from green to yellowish or orange.

Harvesting of Arecanut

Harvesting is done by climbing the tree with the help of ladder and rope ring around the feet. On reaching the top the climber harvest the matured bunch with sharp knife or sickle.

Maturity Signs and Harvesting of Sapota

Maturity Sings of Sapota

1. Fruits generally mature in about 240 to 270 days after flowering.
2. Fruits at full maturity develop a dull orange or potato colour.
3. A mature fruit when scratched slightly with nail shows a yellow streak instead of green streak.
4. Brown scaly material disappears from the fruit surface as the fruits approach full maturity.
5. The milky latex content is reduced.
6. The dried spine like style at the top of the fruit falls or drops off easily when touched.

Harvesting of Sapota

1. The fully matured fruits are harvested with the stalk intact, individually by giving a twist and collected without brushing.
2. Actual Sapota harvester developed by Konkan Krishi Vidyapeeth, Dapoli should be used to avoid post harvest losse.

Maturity Signs and Harvesting of Papaya

Maturity Signs of Papaya

1. Fruits require 125 to 140 days from flowering to maturity.
2. Colour of fruit changes from green to pale green or yellowish.

3. Portion of fruit exposed to sunlight becomes dark yellow in colour.
4. The latex of fruits becomes watery.
5. T.S.S at harvest should be minimum 6%.

Harvesting of Papaya

1. It is harvested by twisting the fruit on dwarf trees, the operation can be done without mechanical aids.
2. Ladders are often used to harvest tall trees. Poles should never be used for harvesting. The peduncle should be trimmed before panicking.

Maturity Signs and Harvesting of Jackfruit

Maturity Signs of Jackfruit

1. The fruit mature in about 180 to 270 days after flowering.
2. Fruit colour changes from green to pale green.
3. A dull, hollow sound is produced when the fruit is tapped by the finger.
4. Fruit spinel become well developed and wide spaced.
5. The last leaf of the peduncle yellows.
6. An aromatic odor develops.
7. The spines yield to moderate pressure.

Harvesting of Jackfruit

1. If the fruit is within the reach, it can be cut with a sickle or twisted unit it snaps off. In tall trees, a sack is placed on the fruit with a rope tied on the peduncle.
2. After cutting, the fruit is gradually lowered to the ground.
3. It is laid on its peduncle against railing for some time to allow the latex to flow and coagulate.

Maturity Signs and Harvesting of Cashew

Maturity Sings of Cashew

1. Fruit matures in about 95 to 120 days from flowering.
2. Depending upon variety apple colour changes from green to yellow or red.
3. Nut becomes hard and ash in colour.

Harvesting of Cashew

1. Harvesting and collection of nuts are done over a period of 10-12 weeks. Where apple is not collected, the fruits may be allowed to fall on the ground and nuts collected periodically during first 4 to 6 weeks. Later most of the nuts are collected in the trees. Towards the end of season nuts can again collected from ground.

Maturity Signs and Harvesting of Ber

Maturity Signs of Ber

1. About 125 to 150 days are required for harvesting of fruits after flowering.
2. Colour changes from green to pale green / red / brownish, yellow, depending upon variety.
3. Fruit become soft.
4. In Umran specific gravity should be less than 1 with golden yellowish colour.

Harvesting of Ber

1. It is harvested from November to March in different parts of the country which is earlier in South and late in North India. The peak season of harvest in northern India is between February- April. The fruit should be picked at right stage of maturity.

Maturity Signs and Harvesting of Guava

Maturity Signs of Guava

1. Fruit mature in 140 to 160 days after flowering.
2. Colour changes from green to pale green or yellow.
3. Fruit becomes soft,
4. A slight depression develops near the stalk end.
5. Ridges on the fruit disappears and it becomes round.

Harvesting of Guava

1. Hand-pulling is employed.
2. Higher branches can be reached with a ladder. The fruit are collected in basket.

3. Picking is done 2 to 3 times a week during harvest season of 8 to 10 weeks.

Maturity Signs and Harvesting of Grapes Maturity Signs of Grapes:

1. A fruit mature in 120 to 135 days, after flowering.
2. Depending upon the variety colour changes from green to yellow on golden yellow or black or pink.
3. being non – climacteric fully ripe fruits are harvested.
4. Easy separation of barriers indicates harvesting stage.
5. Development of characteristics flavours and aroma.
6. *Anabe shahi:* 15 to 16 % T.S.S. Thompson seedless 18 to 20 % T.S.S. Thompson seedless 18 to 20% T.S.S. Harvesting of Grapes: Grapes are harvested when they are fully riped. While harvesting the clusters should be removed from the vine by cutting with a sharp knife near its attachment to the cane. Handle the cluster by the stem.

Maturity Signs and Harvesting of Pomegranate

Maturity Signs of Pomegranate

1. Fruits are ready for harvesting in about 150 to 170 days.
2. Colour changes from green to yellow or yellowish brown.
3. Fruit becomes soft.
4. Ridges develop on the fruits and fruit becomes flat.
5. Floral part dry out.
6. Pressing fruit with finger give cracking or metallic sound.

Harvesting of Pomegranate

The fruits are harvested manually after reaching proper maturity stage. Maturity Signs and Harvesting of Mandarin and Sweet Orange

Maturity Sings of Mandarin and Sweet Orange

1. Fruit mature in about 210 to 240 days.
2. Colour changes from green to pale green or pale yellow or orange.

3. Outer rind looks shiny and one can see oil glands on the fruit.
4. Mandarin 0.4% acidity and 12 to 14 % T.S.S. orange 0.3% acidity and 12% T.S.S.
5. Fruits become soft.

Harvesting of Mandarin and Sweet Orange

1. In India the fruit picker collect the fruits manually by climbing on a ladder when the tree is tall, with a collecting bag on his shoulders.

   In U.S.A mechanical harvesting is done particularly for processing industry.

Maturity Signs and Harvesting of Kagzi Lime

Maturity Signs of Kagzi Lime

1. Flowers mature in 120 to 150 days.

Harvesting of Kagzi Lime

1. In India the fruit picker collect the fruits manually by climbing on a ladder when the tree is tall, with a collecting bag on his shoulders.

   In U.S.A mechanical harvesting is done particularly for processing industry.

Maturity Signs and Harvesting of Pineapple

Maturity Sings of Pineapple

1. Require 105 to 130 days for maturity.
2. Colour of lower portion of fruit turns green yellow.
3. Eyes on the fruit becomes smooth or flat and they buldge on the sides.
4. Eye bracts become loose and turn brown in colour.
5. Mature pineapple should have 12% T.S.S. 0.5 to 0.6% acidity, Specific gravity 0.98 to 1.02.

Harvesting of Pineapple

Harvesting should be done with a sharp knife severing the fruit stalk with a clean cut and retaining 5-7 cm of the stalk.

Maturity Signs and Harvesting of Custard Apple

Maturity Sings of Custard Apple

1. Require 100 to 120 days for full maturity.
2. Scales on fruits becomes prominent, plummy and well space.
3. Fruit become scales or segments turn creamy white.

Harvesting of Custard Apple

Harvesting may extend from September to November depending on the period of flowering. Fruits are harvested manually.

Maturity Signs and Harvesting of Kokum

Maturity Sings of Kokum

1. The fruits generally require 94 days to mature after flowering.
2. The colour of fruit changes from green to light green and red and purplish red when ripe.
3. Physical parameters at mature stage.
   (a) Average weight of fruit 34.01 gm (depends on the type)
   (b) Specific gravity 1.04.
4. Chemical parameters at mature stage. A. T.S.S. 13.68% b. Acidity 3.61 % c. Total sugar 6.35%.

Harvesting of Kokum

1. Generally riped fruits (red to purplish red colour) are harvested. The fruits are harvested manually by picking or beating with MANGA. The "Atul Sapota" Harvester " developed by Konkan Krishi Vidyappeth , Dapoli can be used for harvesting kokum fruits.

Maturity Signs and Harvesting of Jamun

Maturity Sings of Jamun

1. The seedling Jamum plants starts bearing after 8-10 years of planting, while grafted once after 6-7 years.
2. Colour changes from green to deep purple or black with full size.

Harvesting of Jamun

1. The fruits of Jamun are generally harvested daily and sent to the market on the same day (shelf life 2-3 days). The ripe fruits are picked singly by hand. The picker climbs the trees, with bag of cotton slung on the shoulder and harvests the ripped fruits.

Maturity Signs and Harvesting of Karonda

Maturity Sings

1. Fruit generally require 87 to 90 days to mature after flowering.
2. Colour changes from light green to dark green.
3. Red blush develops on the fruits of some ypes.
4. Specific gravity just above 1 (1.012).
5. T.S.S. 13.5%. Total sugar 2.35% and Acidity 2.62%.

Harvesting

1. The fully mature fruits are harvested with the stalk in act.

Maturity Signs and Harvesting of Cabbage, Cauliflower and Knol-Khol

Maturity Signs of Cabbage

1. At low elevations cabbage varieties mature in about 62 to 110 days from field setting and about 81 to 125 days at high elevations.
2. Solidity and firmness of head are the usual maturity characteristics used.
3. Head colour turns a lighter shade of green when full development is a attained.
4. They have tendency to burst or loosen the leaves beyond the marketable stage.

Harvesting of Cabbage: The heads are cut with knife frequently attached with some wrapper leaves.

Maturity Signs of Cauliflower

1. The best stage of maturity is determined by curd size and condition. Local growers usually harvest the head

upon the desired size and before the curds become discoloured, loose or otherwise blemished.

2. The head should be compact and not to be broken into segments.
3. Over mature head, becomes too long, flower stalks elongates resulting in loose, leafy condition and possess poor market value.

Harvesting of Cauliflower

The curds are cut off the stalk with large sharp knife. The large leaves are trimmed away until sufficient jacket leaves remain to protect the curd.

Maturity Signs of Knol-Khol

The fruits are harvested manually after reaching proper

1. Harvesting is done when it attains its marketable size i.e. 5 to 7 cm in diametres and bright colour depending upon variety.
2. At this stage the edible portion is tender and non-fibrous.

Harvesting of Knol-Khol

The knobs are harvested with sharp knife, when they are mature. Foliage is kept while harvesting.

Maturity Signs and Harvesting of Tomato and Chilli

Maturity Signs of Tomato

1. Harvesting depends upon the purpose for which they are used. Four maturity stages are generally recognized.
   (a) *Green stage:* The mature green fruits are generally harvested to send them to the distant market.
   (b) *Pink Stage:* At this stage colour turns to pink or red at the blossom end. They are picked for local market.
   (c) *Ripe stage:* At this stage surface of the most of the fruits is red and the soften of fruits begins.
   (d) *Fully Ripe:* At this stage fruits have approached maximum colour development and are soft. Starch is charged into sugars. They are generally consumed or used for canning and processing.

Harvesting of Tomato

Depending upon the purpose for which they are used and market distance, tomatoes are harvested manually by plucking the fruits at different maturity stages.

Maturity Sings of Chilli

1. Chillies are harvested at two stages, one for green vegetables and the other as dry chillies.
2. Green Chillies are harvested when they are fully mature and before they change from green to red.
3. Chillies for drying should be harvested when colour changes from green to red.

Harvesting of Chilli

Depending upon the purpose for which chillies are to be used, fruits are picked either green or fully red ripe. Maturity Signs and Harvesting of Bell Pepper

Maturity Signs of Bell Pepper

1. Generally bell peppers are harvested in the mature green stage. They should be of good shape, waxy, firm and shiny.
2. Peppers for canning are allowed to attain full size and develop a red colour before harvesting.

Harvesting of Bell Pepper

The sweet peppers are picked with an upward twist which leaves a piece of stem attached.

Maturity Signs and Harvesting of Brinjal

Maturity Signs of Brinjal

1. The fruits must be harvested as soon as they have attained the desired size, colour and before they lose their bright glossy appearances and become dull.
2. The fruits are edible from the time they are quarter growth until they are near ripe.

Harvesting of Brinjal

The fruits are harvested from the plant by cutting with a small pruning shears or a sharp knife. The fleshy calyx and a short piece of stalk are left attached to the fruit.

Maturity Signs and Harvesting of Carrot and Radish

Maturity Sings of Carrot

1. Depending upon variety carrot become ready for harvesting within 100 to 120 days from sowing.
2. Depending in the variety, the desired size is the primary consideration in harvesting the roots.
3. Oversized roots are not acceptable.
4. Generally small to medium sized carrots are preferred.
5. At marketable stage carrots should have at least 2.5 to 4 cm diametre at the upper end.

Harvesting of Carrot

Early carrots for the market are pulled out when they partly developed. They are normally du out, when the soil is sufficient moist with spade or a khurpi. The roots are trimmed and washed before sending them to the market.

Maturity Sings of Radish

1. European type is ready for harvesting within 30 days from sowing and Asiatic type within 45 to 60 days from sowing.
2. At this stage roots are mild, tender and crisp and usually of the proper marketable size.
3. Roots must be harvested before they become pithy, bitter and fibrous.

Harvesting of Radish

In India, harvesting is done manually. A light irrigation may be given before harvesting to facilitate lifting of roots. The roots are washed, graded and tied in bunches. In advanced countries, commercial radish growers use a single row harvester that pulls the plants from the soil.

Maturity Signs and Harvesting of Onion and Garlic

Maturity Signs of Onion

1. Harvesting depends upon the purpose for which the crop is planted and maturity will depend upon the variety being used.

2. *Green Bunch Onion:* Green onion is best when they are of lead pencil unit a small bulb is formed.
3. *Ripe Bulbs:* Ripe bulb crop is ready for harvest in 3 to 4 months after transplanting. Neck fall is the indication of maturity.

   The best time to harvest onion is when 60 to 70% of tops have broken over. Development of red pigment and the characteristics pregnancy of the variety are also important harvest indices.

Harvesting of Onion

If the bulbs are mature, they are pulled out easily by hand if the soil is not hard or compact, otherwise, they may be dug out with a shovel or khurpi.

Maturity Signs of Garlic

1. The crop may be harvested when the top turns yellowish or brownish and shows sings of drying up.
2. Crop gets ready for harvest in about 100 to 140 days after planting.

Harvesting of Garlic

The bulbs lifted manually, cleaned and the leaves are tied at the top. The bulbs are dried for a week or so under shade.

Maturity Signs and Harvesting of Bhendi and Pea

Maturity Sings of Bhendi

1. Fruits are harvested when the pods are still mature, young and tender. They are harvested at an interval of 2 to 3 days.
2. The best time of harvest being 6 to 7 days after opening of flower.
3. Mature pods are fibrous, tough, and unit for human consumption.

Harvesting of Bhendi

The tender young fruits 7 to 10 cm long should be harvested every alternate day. Frequent picking promote fruit development and enhances yield.

Maturity Signs of Pea

1. Early varieties require 45 to 60 days and late varieties require 70 to 100 days from sowing.
2. Peas must be picked up at the proper stage of maturity, because they start loosing their rapidly after reaching the edible stage.
3. Quality depends upon sugar content and tenderness. With increasing maturity and size , sugar content declines rapidly, with a correspondence increase in starch and protein.
4. High sugar content is indication of high quality.
5. Harvest the pods when colour of pods changes from dark green to light green with good size.

Harvesting of Pea:

The pods should be harvested at proper maturity by picking at an interval of 12 to 15 days. Picking daily will injure the plant.

Maturity Signs and Harvesting of Potato and Cucumber

Maturity Sings of Potato

1. Potatoes are harvested when they attain sufficient size. Early varieties 57 to 100 days, late varieties 120- 160 days.
2. Skin slipping from the tuber, starch content and leaf senescence or top drying are the harvest indices.

Harvesting of Potato

Care should be taken to save the tuber from injury and immediately after harvest they should be left in the scorching sun. otherwise they develop sunscalds. Digging is done with spade or other hand tool. Suitable tractor operated potato diggers have been developed now. There should be optimum moisture in the soil at the time of harvest.

Maturity Sings of Cucumber

1. Fruits are ready for harvest 45 to 55 days after sowing.
2. Harvest immature fruits.
3. In cucumber the proper stage of maturity is judged by size and not by the age of the fruit.

4. Cucumber for slicing should be picked when they are 15 to 35 cm long, whereas for pickling 6 to 15 cm long. In case of slices at marketable stage, spines on fruit becomes soft and fall down.
5. In general cucumber may be picked at any stage of fruit growth, provided yellowish has not started.

Harvesting of Cucumber

The fruit should be picked at frequent intervals in order to avoid losses due to over sized or over mature fruits. They are generally picked at 2 to 4 days intervals depending on weather.

Maturity Signs and Harvesting of Water Melon and Muskmelon

Maturity Signs of Water Melon

1. Size of the fruits and colour of the rind are not good indicators.
2. Fruits are ready for harvest in 90 to 120 days from sowing.
3. Four criteria are commonly used in determining the maturity of the fruits:
    a) *Withering of tendril:* The tendril accompanying the fruit withers as fruit ripeness.
    b) *Thumping:* Ripe fruits when thumped with finger give out heavy dull sound, whereas the immature fruits give metallic sound.
    c) The portion of fruit, which rests on the ground, turned yellowing at maturity.
    d) Ripe fruits produce a crisp, cracking noise on being pressed with the flat of the hand.

Harvesting of Water Melon

The well mature fruits are harvested manually with the help of sharp knife or by twisting.

Maturity Signs of Muskmelon

1. Fruits will be ready for picking in about 110 days depending upon variety. There are two groups of

cultivars which behave distinctly. In one group, the fruits when mature slips out easily from the vine with little pressure or jerk or if not it will remain separated from the vine next day. This is called full slip stage. In some Indian cultivars, green stripes on the skin begin to turn yellow during maturity.

Harvesting of Muskmelon

Muskmelon is a climacteric fruit which ripens during transit and storage and hence, it is harvested before it is fully ripe.

Maturity Signs and Harvesting of Red Pumpkin and Gourds

Maturity Sings of Red Pumpkin

1. Fruits require 110 to 120 days for maturity from sowing.
2. Fruits should be harvested at full stage of maturity when the rind has hardened.
3. Colour changes from green to yellow or orange. Some varieties remain green only after full maturity.

Harvesting of Red Pumpkin

The fruits are harvested manually by keeping a portion of stem attached. The fruits are placed in heap and allowed to cure for a few days.

Maturity Signs of Gourds

1. Tender fruits are harvested.
2. Proper stage of maturity in bottle gourd, sponge gourd and ridge is judged by
    a) Its fruit size
    b) The gentle press of finger nails which easily penetrate the epidermis.
    c) Plugging the fruits which show fine tender flesh and quite immature seeds. However, this test makes the fruit unfit for market.
3. In most cases fruits require 12 to 15 days after pollination to attain marketable size.

Harvesting of Gourds

Harvesting is done at short intervals. The fruits are harvested manually by keeping a portion of stem attached.

Maturity Signs and Harvesting of Sweet Potato

Maturity Sings of Sweet Potato

1. They are harvested 3.5 to 4 months after planting.
2. Leaves start yellowing
3. After cutting the tuber if white gum remains white, then they are ready for harvest , it it turns blackish on green then tubers are said to be immature.

Harvesting of Sweet Potato

The crop is harvested by digging with spade or kudali.

Maturity Signs and Harvesting of Leafy Vegetables (Spinach, Amaranthus and Fenugreek)

Maturity Signs

1. They are harvested 3 to 4 weeks after sowing,
2. Harvesting should be done, when the leaves are immature and tender, but large enough.
3. Old leaves are bitter and unfit for consumption.

Harvesting

Spinach: First cutting 3 to 4 weeks after sowing. Leaves are cut manually from the base. It gives 3 to 4 pickings at 15 days interval.

Amaranthus: Uprooting is done 3-4 weeks after sowing. If clippings are to be taken then first clipping 25 to 30 days after sowing and other at 20 to 25 days interval. Total 2-4 clippings.

Fenugreek: When common methi is taken, the younger shoots are nipped off in about 3 weeks after sowing. Later on the whole plant is often pulled out, bunched and marketed. The kasuri methi gives a number of cutting. First cutting 25-30 days after sowing. Subsequent cuttings taken at 12 to 15 days interval. After 2-3 cuttings plants are left for seeding. Pre Harvest Factors Responsible for Maturity, Ripening and Deterioration of Horticultural Produce.

Post harvest losses in fruits and vegetables are very high (2-40%). About 10-15 % fresh fruits and vegetables shrivel and stale, lowering their market value and consumer acceptability. Minimizing these losses can increase their supply without bringing additional land under cultivation. It will also help keep pollution under control. Improper handling and storage cause physical damages due to tissue breakdown. Mechanical losses include bruising, cracking , cuts, microbial, spoilage by fungi and bacteria, whereas physiological losses include changes in respiration , transpiration, pigments , organic acids and flavour.

About 36% of vegetables decay due to soft-rot bacteria, whereas 30% of fruits decay due to Penicillium species. The losses can be minimized by proper pre and post harvest treatments.

Pre-Harvest Factors

Selection of Varieties: Varieties with higher yield, better keeping quality, slower ripening and longer shelf-life under ambient condition and better processing quality should be developed and commercially grown.

Cultural Operations: Cultural operations help prolong shelf-life of fruits and vegetables. Pruning or thinning increase the fruit size and decrease TSS and acidity. The postharvest quality of citrus fruits increase significantly when Trifoliate orange, tangelo or Cleopatra are used as root stock. Quality of fruits is improved by the application of K, Mg and Zn , while higher N and P deteriorate the quality. Excessive irrigation before harvesting decreases the shelf-life and sensory quality, while insufficient irrigation enhances the maturity of the crop.

For root crops, preparation of the soil to a tilth of porous nature is necessary to avoid root forking. Irregular irrigation causes cracking of carrot and radish and splitting of outer scales of onions.

In onion and garlic, irrigation should be stopped 3 weeks before harvesting to ensure better keeping quality. Heavy application of nitrogenous fertilizers causes faster tissue deterioration, while essential supply of 'Kim-proves' keeping quality of vegetables. Molybdenum –efficiency in cabbage cause heart –rot, manganese deficiency in peas leads to marsh spots and excessive irrigation and fertilization cause hollow heart in potato.

Pre-Harvest Treatments: Postharvest shelf-life of fruits and vegetables is improved by pre-harvest application of chemicals. Three spray of Topsin-M (0.1%) or Bavistin (0.1%) at 15 days interval before harvesting can control anthracnose and stem – end rot in mango. Similarly, postharvest decay of Nagpur mandarins can be controlled by 3 pre-harvest sprays of 0.1% Ben late or 0.1% Topsin-M or 0.1 Bavistin at 15 days interval. Pre-harvest application of maleic hydrazide reduces sprouting of onions and potatoes during storage. In rabi and kharif onions, application of 1500-2000 ppm maleic hydrazide, 75-90 days after transplanting reduces sprouting during 4-5 months of storage in ventilated structures. Postharvest diseases of tomato and onion can be controlled by 3 pre-harvest sprays of 0.2 Difolatan at 10 days interval. Similarly pre-harvest application of growth promoters such as N-benzyladenine (10-20 ppm) prolongs shelf-life of leafy vegetables.

Maturity: The postharvest quality and storage life of fruits are controlled by maturity. If fruits are harvested at a proper stage of maturity their quality is excellent. Vegetables are harvested as and when they attain maximum size and yet are tender. Over maturity in root crops causes sponginess and pithiness. Their harvesting should not be delayed. Delay in harvesting and onion and garlic reduces their storage quality.

Harvesting: Harvesting should be done during cooler part of the day preferably in the morning. The produce should be shifted to shade as early as possible. Harvesting during hot periods raises field heat of the produce, causing wilting and shrivelling. Harvesting during or immediately after rains should not be carried out since it creates most favourable conditions for multiplication of microorganisms. Citrus fruits becomes susceptible to damage if harvested during rains as their rind becomes turgid and prone to easy brushing, sunscald and oleocellosis. Care in harvesting is necessary as any bruises or injuries during harvesting may later manifest as black or brown patches making them unattractive. Latex coming out of stem in mango should not be allowed to fall on fruits as it creates a black spot. Injury to peel may become an entry point for micro-organisms, causing rotting. Some harvesting gadgets have been developed. E.g. mango harvester in Lucknow- (CSH), Bangalore (IHR) and Ratnagiri (KKV). These need to be popularized.

Many fruits and vegetables are harvested unripe for their safe handling. Transportation and marketing but they must be matured when harvested so that they can ripen later on normally and develop good eating quality. Post Harvest Factors Responsible for Maturity, Ripening and Deterioration of Horticultural Produce

Curing: Curing is conducted immediately after harvesting. It strengthens the skin. The process is induced at relatively higher temperature and humidity involving suberization of outer tissues followed by the development of wound periderm which acts as an effective barrier against infection and water loss.

It is favoured by high temperature and high humidity. Potato, sweet potato, colocasia, onion and garlic are cured prior to storage or marketing.

In sweet potato, this condition is most rapid at 33 0C and relative humidity of 95%. Potato tubers are held at 18 0C for 2 days and then at 7 0C – 10 0C for 10-12 days at 90 % relative humidity. Curing also reduced the moisture content especially in onion and garlic. Drying of superficial leaves of onion bulbs protects them from microbial infection in storage. Maximum safe temperature for onion curing at field is 37.8 0C for 3-5 days. Artificial curing of onions in crates at 40 0C for 16 hours reduces rot losses in storage.

Degreening: Degreening is the process of decomposing green pigments in fruits usually by applying ethylene or other similar metabolic inducers to give a fruit its characteristic colour as preferred by consumers. It is applicable to banana, mango, citrus, and tomato. The time required to degree a fruit depends upon the degree of natural colour break and maturity. The higher the green colour and more mature a fruit is the less time is required to reduce the chlorophyll to a desired level.

Degreening is carried out in special treating rooms with controlled temperature and humidity in which low concentration of ethylene (20ppm) is applied. The ethylene should be supplied from a gas cylinders.

These rooms are thoroughly ventilated to keep the Co2 level below 1% which does not allow higher colouring. If kerosene fumes are placed outside the degreening room , they enter the room through ducts by forced ventilation. Despite the fire hazard involved, kerosene fumes produce better coloured fruits than pure ethylene. It is due to good ventilation. The best Degreening temperature is 27 0C. Higher temperatures delay Degreening. The relative humidity should be 85-90%. Higher humidity levels cause condensation during Degreening and are associated with slow Degreening and increase in decay. Low humidity though checks decay, causes excessive shrinkage , shrivelling and peel break down. In another method, frits in containers are sealed by 2 sheets of plastic film and water. the PVC (0.2mm thick) can also be used. The ethylene is introduced from a can of 4.2 litres capacity in the film which can cover 1.2 tonnes of fruits, the ethylene concentration becomes nearly 1000 ppm resulting in satisfactory colouring. After 15 hours the film is removed enforcing the fruit to air. Degreening takes 3 days. Ethylene accelerates decomposition of chlorophyll without significant affecting the synthesis of carotenoid pigments.

Pre-Cooling: High temperatures are detrimental to keeping quality of fruits and vegetables, especially when harvesting is done during hot days. Pre-cooling is a means of removing the field heat. It slows down the respiration of the produce, ,minimizes susceptibility to attack of microorganism , reduces water loss and eases the load on cooling system of storage or transport. Peas and okra which deteriorate fast, need to prompt cooling. Sometimes stage of ripening and level of field heat of produce also determine the need for pre-cooling. For example, unless tomatoes are avove 26.7 0C and ripening and level of field heat of produce also determine the need for precolling. For example , unless tomatoes are above 26.7 0C and ripening is to be delayed there is no need for pre-cooling.

In air cooling, cool air can be obtained from cold storage. Temperature should not be less than -10C to avoid freezing. Where night temperatures are low, doors of the store rooms can be opened for cooling in the night.

In water cooling, field heat is removed quickly. It is used for leafy vegetables to retain their texture and freshness. Ice can be added to bring down the temperature. However, temperature should be controlled to avoid chilling injury in cold sensitive fruits and vegetables. Hydro cooling at 12 0C to 15 0C with 500 ppm Bavistin increases shelf-life of mango. In Alphonso mango, it also reduces the incidence of spongy tissue.

Washing and Drying: Most of the fruits and vegetables are washed after harvesting to improve their appearances prevent wilting and remove primary inculum load of microorganisms. Hence a fungicide should be used in washing water. Washing improves shelf-life of bananas by delaying their ripening. After washing, excess should be removed which would other encourage microbial spoilage. Root and tuber crops are often washed to remove the soil adhering to these.

Sorting and Grading: Immature diseased and badly bruised fruits and vegetables are sorted out, most of the countries have their own set of standards of domestic trade and for international trade, standards have also been defined. Grades are based on size, weight, colour and shape. Grading is done manually or mechanically.

Disinfestations: Papaya, mango, melon and other fruits are susceptible to fruit fly attacks. Disinfestations is done either by vapour heat treatment at 43 0C with air saturated with later vapour for 6-8 hours by ethylene dibromide fumigation (18-22 g of EBD/ cubic metre for 2-4 hours. Residues of inorganic bromide must not exceed to 10vg/g) or by cold treatment (exposure of fruits to near freezing temperature for a specified period).

Post-Harvest Treatment: Post –harvest application of Bavistin (0.1%) and Topsin (0.1%) controls storage disease in mango. In Nagpur mandarins, hot water treatment with Imazalil (0.1%), Bavistin (0.1%) and Benlate (0.1%) is most effective. A complete inhibition of sprouting of cool chamber (evaporatively cooled) stored potatoes for 4 months and 5 months is achieved by spraying them with an aqueous emulsion of CIPC @ 50 mg/ kg of tubers respectively before completion of dormancy period.

Waxing: Fruits and vegetables have a natural waxy layer on their outer surface which is partly removed by washing. An extra discontinuous layer of wax applied artificially with sufficient thickness and consistency to prevent anaerobic condition within the fruits provides necessary protection against decay organisms.

Waxing is especially important if tiny injuries and scratches on their surface are present. These can be sealed by wax. Waxing also enhances the gloss of fruits or vegetables. Therefore, appearance is improved, making them more acceptable.

If refrigerated storage facilities are not available , protective skin coating with wax increases the storage life of fresh fruits and vegetables at ambient temperatures. There are two types of wax emulsion: Wax 'W' and Wax 'O'. The composition 'W' does not impart any gloss to fruits and vegetables and hence where gloss is required for improving marketability of the produce, composition wax 'O' is recommended. Both these emulsion contain 12 % total solids. The application of wax emulsion to freshly harvested healthy produce protects them against excessive moisture loss, higher rate of respiration, heat buildup or thermal decomposition. Texture and quality of the fresh produce is maintained as near the fresh condition as possible for a long time. The wax emulsion without fungicide does not protect fruits and vegetables against microbial spoilage. Therefore, suitable fungicides are added to the wax emulsion.

Ripening of Fruits: Ripening transforms a physically mature but inedible plant organ into a visually attractive taste and smell sensation. It marks the completion of development and commencement of senescence with life of a fruit and is normally an irreversible event. Ripening can be achieved by the application of ethylene. Accurate quantity of ethylene should be used in the ripening room at regular interval. A concentration of CO2 about 1% delays ripening. Hence, through ventilation Is essential) marketing is alkaline using caustic soda.

Pre-Packaging in Plastic Film: Pre-packaging increases the shelf life by create a modified atmosphere with an increases in concentration of Co2 in the package. The packaging material used should provide reasonable access to oxygen. For this, beginning films like polystyrene and cellulose acetate are used. But together LDPE films which have high O2 and Co2 transpiration rates are more durable, the pouches must have perforation to transmit oxygen and carbon dioxide rapidly enough for the respiration of fresh produce. The pouch used reduces bruising facilitates inspection, reduces moisture loss and prevents dehydration. It also creates modified atmosphere. In pre-packaging leaves, stalk, stem, etc. are trimmed washed cleaned and weighted quantities are put in pouches. Ethylene absorbents may be added to the package wherever required to retard the ripening process. Hydrated lime inserts may also be beneficial in controlling CO2 concentration within the film package.

Bibliography

Ainsworth GC.: *Introduction to the History of Mycology*, Cambridge University Press, Cambridge, UK, 2016.

Alexopoulos CJ; Mims CW; Blackwell M.: *Introductory Mycology*, John Wiley and Sons, UK, 2016.

Collmer A.; Davis RE; Gillaspie AE; Civerolo EL: *Plant Pathogenic Bacteria, Dordrecht*, Kluwer Academic Publishers, The Netherlands, 2017.

Cooper Jim: *Bacterial Plant Pathogens and Symptomology*, WSU County Extension, SJI, 2006.

Crosa JH; Kado CI: *Molecular Mechanisms of Bacterial Virulence*, Dordrecht, Kluwer Academic Publishers, Crosse JE, The Netherlands, 2016.

Deacon J.: *Fungal Biology*, Blackwell Publishers, Cambridge, Massachusetts, 2005.

Dowson WJ: *Plant Diseases due to Bacteria*, Cambridge University Press, Cambridge, UK, 2017.

Fahy PC; Persley GJ: *Plant Bacterial Diseases: A Diagnostic Guide*, Academic Press, New York, USA, 2020.

Farrand SK: *Agrobacterium Radiobacter K84: A Model Biocontrol System*, Academic Press,New York, USA, 2020.

Fokema NN: Van den Heuvel J, *Microbiology of the Phyllosphere*, Cambridge University Press, Cambridge, UK, 2019.

Fokkema N.: *Microbiology of the Phyllosphere*, Cambridge University Press, Cambridge, UK, 2017.

Goodman RN: *Encyclopedia of Plant and Crop Science*, Springer Publishing, New York, 2017.

Goto M: *Fundamentals of Bacterial Plant Pathology*, Academic Press, New York, 2022.

Hall IR.: *Edible and Poisonous Mushrooms of the World*, Timber Press, Portland, 2003.

Hayward AC; Hartman GL: *The Disease and its Causative Agent*, CAB International, Wallingford, UK, 2014.

Hayward AC; Waterston JM: *CMI Descriptions of Pathogenic Fungi and Bacteria*, Wallingford, UK: CAB International, 2015.

Jackson RW: *Plant Pathogenic Bacteria: Genomics and Molecular Biology*, Caister Academic Press, 2009.

Kirk PM; Cannon PF; Minter DW; Stalpers JA.: *Dictionary of the Fungi,* CAB International, Wallingford, UK, 2008.

Lee I-M; Davis RE; Dawn E.: *Phytoplasma: phytopathogenic Mollicutes*, APS Press, St Paul, Minnesota, USA, 2020.

Lee RE; Warren GJ: Gusta LV: *Biological Ice Nucleation and Its Applications*, APS Press, St Paul, Minnesota, USA, 2015.

Lelliot RA; Stead DE: *Methods for the Diagnosis of Bacterial Diseases of Plants*, Blackwell, Oxford, UK, 2017.

Lindow SE; Brandl MT: *Microbiology of the phyllosphere*, Cambridge University Press, Cambridge, UK, 2021.

Moore, David: *Fungal Conservation: Issues and Solutions,* Cambridge University Press, 2021.

Priest FG; Ramos-Cormenzana A.: Tindall B, *Bacterial Diversity and Systematics*, Plenum Press, New York, USA, 2019.

Prior P.; Allen C.; Elphinstone J.: *Molecular and Ecological Aspects*, Springer Publishing, Berlin, Germany, 2014.

Schaechter M; Summers WC.: *Encyclopedia of Microbiology,* Academic Press, San Diego, USA, 2020.

Taylor EL; Taylor TN.: *The Biology and Evolution of Fossil Plants*, Prentice Hall, Englewood Cliffs, New Jersey, 2013.

Wang Daowen; Maule Andrew J.: *A Model for Seed Transmission of a Plant Virus*m, University of Missouri Press, USA, 2016.

Zaitlin, Milton and Peter Palukaitis: *Advances in Understanding Plant Viruses and Virus Diseases*, Academic Press, New York, USA, 2020.

Index

❑❑❑